Presenting Data

Presenting Data

How to Communicate Your Message Effectively

Ed Swires-Hennessy

Retired Government Statistician, UK

This edition first published 2014
© 2014 John Wiley & Sons, Ltd

Registered office
John Wiley & Sons Ltd, The Atrium, Southern Gate, Chichester, West Sussex, PO19 8SQ, United Kingdom

For details of our global editorial offices, for customer services and for information about how to apply for permission to reuse the copyright material in this book please see our website at www.wiley.com.

Wiley also publishes its books in a variety of electronic formats. Some content that appears in print may not be available in electronic books.

Designations used by companies to distinguish their products are often claimed as trademarks. All brand names and product names used in this book are trade names, service marks, trademarks or registered trademarks of their respective owners. The publisher is not associated with any product or vendor mentioned in this book.

Limit of Liability/Disclaimer of Warranty: While the publisher and author have used their best efforts in preparing this book, they make no representations or warranties with respect to the accuracy or completeness of the contents of this book and specifically disclaim any implied warranties of merchantability or fitness for a particular purpose. It is sold on the understanding that the publisher is not engaged in rendering professional services and neither the publisher nor the author shall be liable for damages arising herefrom. If professional advice or other expert assistance is required, the services of a competent professional should be sought.

Library of Congress Cataloging-in-Publication Data

Swires-Hennessy, Ed.
    Presenting data : how to communicate your message effectively / Ed Swires-Hennessy.
        pages cm
    Includes bibliographical references.
    ISBN 978-1-118-48959-8 (hardback)
    1. Communication of technical information.   2. Statistics–Charts, diagrams, etc..   3. Communication in science.   I. Title.
    T10.5.S95 2014
    001.4'226–dc23

                                                                                2014014248

A catalogue record for this book is available from the British Library.

ISBN: 978-1-118-48959-8

Set in 10/12pt Times by Aptara Inc., New Delhi, India

1   2014

Contents

List of Tables

List of Figures

Introduction

Data are everywhere! Data provide the vital evidence to develop modern society and there has been a boom in the number of people and organisations collecting, handling and trying to interpret data. However, many do not recognise the issues of presenting data. The result is that there is often poor public understanding of the data and statistics presented.

It is most important that the handling, interpretation and communication of data and the statistics derived from them are made as easy as possible for the wide and diverse audience of people who try to understand and use statistics and data.

The importance of making presentation as clear as possible has been magnified since the Internet has made data and statistics even more widely available to audiences for whom they might not originally have been intended. There is no hiding the fact that the data and statistical skills of the majority of the population are weak. Therefore, the more help that is given through good presentation, the better chance that the message will be understood. It must not be assumed that, without the help of clear presentation, the audience will be able quickly to grasp the message, meaning or relevance of the data or statistics.

The aim of this book is to look at numerical information through the eye of the audience or user. It builds on the fundamentals of numeracy that are taught in schools to all and examines cases where the interpretation would be helped by better presentation.

The subject of this book rarely appears on any university or college course and hence the principles of effective presentation are virtually unknown. This results in ineffectively and poorly presented numerical information which can mislead the audiences and could lead to incorrect decisions being made.

Current society is shaped by decisions based on evidence driven by data. At the least, misinformation gathered from ineffectively and poorly presented numerical information can lead to the main and fundamental message of the information being lost with time wasted in seeking to identify what the presenter is trying to communicate. Wrong decisions taken on poor or poorly presented data could be very costly.

Conversely, efficiently and well-presented numerical information can be assimilated quickly and accurately leading to a better and correct understanding of the message from the information. Linked to decisions, such information will form the justification and rationale.

Throughout the book, four basic principles are applied – all beginning with the letter C.

> ➤ Correct. The assumption is that all of the data being presented has been checked by the producer before dissemination. Examples are given where this is not the case.

> ➤ Clear. The presentation should always have a specific aim and the resulting presentation of information should meet the aim – whether in tables, charts or text.

> ➤ Concise. The producer is often tempted to present more information than is actually necessary. This may be the result of pride that a great deal of effort has been expended in the collection of detailed data and thus should be shown. Detail here could either be more numbers than necessary to communicate the message or the inclusion of more digits in numbers than required to distinguish differences. However, such presentation can often obfuscate the basic message.

> ➤ Consistent. Presented information should be consistent in units shown when asking a user to compare data (don't ask a user to compare a number rounded to thousands with a number rounded to millions). Further, when abbreviating words, the same abbreviation should be used throughout an organisation – not just within a report. One example quoted in Chapter 4 shows the formatting of financial year in three different ways in the same paragraph! Consistency also should be seen in the presentation of data: if a country's national statistical institute uses a full stop for the thousands separator and a comma for the decimal separator, this should be the case for all of the information presented by that office. From a table of seven indicators for one country[i], the following was extracted:

Indicators	
The rate of change in the Consumer Price Index (CPI)	−0,9 %
The rate of change in the index of industrial production (vol)	5.1%
The rate of change in the index of industrial production (vol)	−1.9%

The first figure used a comma for the decimal separator and the other two a full stop. Further, the first figure had a space after the number and before the '%' symbol. The table was mixed numbers and percentage data so the percentage symbols

---

[i] http://www.bhas.ba/?lang=en on 30 September 2013.

were appropriate. However, the thousands separator used in the table was ... a full stop – the same as the decimal separator in two of the figures above.

When presenting information on the Internet, a fifth word is also appropriate.

> Current. Nothing frustrates a user more than having out-of-date messages on websites, for example, a message on 1 September highlighting that a conference is forthcoming in July of that year. An example seen recently during a review of a website (in September 2013) was:

You can also access all completed parts of ÖNACE 2008, which is due to come into effect in Austria in 2008.

This does not mean that we should not keep historical documents on our websites but just that the basic presentation and referencing are up to date.

The book comes with a health warning: the reader will not look at tables and charts in the same way ever again! The hope, of course, is that the result will be better presentation of numerical information with greater understanding and communication for the user.

Ed Swires-Hennessy
Newport, UK
March 2014

Preface

This book is based on 40 years' worth of experience presenting data – and their messages – to policy colleagues and other users of government statistics. The experience formed the basis of a training course I have run across many organisations, inside and outside government, over the last 30 years. More recently, this course has been run many times in the Royal Statistical Society's training programme. The messages for presenters of data which are included here are appropriate for all types of presentation including data visualisations and infographics. If the fundamental principles are not observed, the audience for the presentation will have to work much harder than necessary to derive the message.

The book is intended to be read by all who present data in any form and the chapters are structured so that they are independent of each other. Poor presentation is everywhere. Basic principles are forgotten or ignored. The result is that audiences are presented with confusing and poor tables and charts and are asked to do calculations in their heads to understand the information.

Throughout, I have looked at the presented data in whatever form through the eyes of a user – essentially the audience for what is presented. By getting producers to perceive the data as a user means that a producer of data will not present users with what they have but what they need in a form that any message can easily be acquired.

The book is for information providers. Throughout, the word 'data' is used as a plural word: the singular would be datum.

Audience

The book will be essential for anyone who needs to present numerical information including:

- Administrators and managers – who present numerical information to colleagues
- Journalists and the associated graphic designers
- Researchers
- Social scientists
- Statisticians

- Economists

- Scientists and specialists in other numerate disciplines

Essentially, it will be of use to anyone who is trying to communicate a message with numerical content. Only basic numeracy is assumed – and some of the basics are restated so that even those would not be taken as given. Hence this book is applicable to all in the university system – from first year to PhD students, to those in a country's administration – from the lowest to the highest level and for all levels of researcher – particularly those who have to present any numerical information. Within the private sector, the book is applicable to all levels of management and others who communicate numerical information (sales information, performance measures, product information, process information): again this is across all levels of the sector.

This book does not include complex statistical formulae or concepts (e.g. probability, statistical significance) but does include a couple of simple formulae (e.g. area of a rectangle and area of a circle). These are used to illustrate the rationale for some of the diagrammatic presentations of the data.

How to read

You can take this book as a guide and read through the whole or use it as a reference text. For both, I suggest it is essential to read the Introduction and Chapter 1 first as these set out the fundamental principles of the presentation of data.

Chapter 1 looks at how all of us in the Western world understand number and reminds all of the basic elements of presentation that were taught in primary schools – to us and our users. Consideration is given to how data should be rounded. By taking away digits, many will argue that a loss of precision has occurred: that is true – but the increase in understanding and communication of what is happening more than outweighs this.

Chapter 2 considers the elements of both reference and summary tables and how they are read. Messages are generally associated with summary tables and examples are given of how users seek to view and understand what is presented.

Chapter 3 shows the types of chart that can represent the data and gives many examples of how charts can be improved to help with message delivery. Knowing how the charts are viewed should assist in the design of all future charts. Issues of each type of chart are discussed and principles developed.

Chapter 4 deals with the fundamentals of including numbers in text and what needs to be done to have the numbers understood quickly. When much data is available, sorting out the key messages from them is essential for good communication as is the language used to describe them.

Chapter 5 moves into the area of Internet presentation of data. The Internet tools for presenting data have been slow to develop but are now allowing users to harness the power of the Internet to derive messages themselves.

Acknowledgements

Many friends over the years have encouraged me to collect examples and teach the basic principles of presentation. In particular, Mike LeGood from the then Civil Service College and Inge Feldbaek from Denmark Statistics were instrumental in encouraging me in the national and international contexts respectively.

Sarah Jones and her team assisted with the first attempt to summarise the course whilst I worked in the Local Government Data Unit – Wales. Alan Smith OBE, from the Office for National Statistics, United Kingdom, has for many years been a great encourager and a provider of examples. Many other colleagues over the years have assisted by providing examples for use in teaching.

Particular thanks are due to John Brettell, who assiduously waded through proofs to find the minutest of issues, to Sarah Jones and David Marder for their insightful reviews of various chapters and to Richard Davies, the Project Editor at Wiley, who has tried to manage me.

1

Understanding number

This chapter deals with the basics of how users perceive numbers and how they should be presented to allow maximum uptake of the meaning of the data.

A false assumption of many who seek to communicate numerical information is that their audience is as able to handle the information as they are. The available data in for the majority of OECD countries is that numeracy skills are significantly below those for literacy.

Numeracy skills in the general population in England are poor – and are not improving. Information from a skills survey in England in 2011[1] noted that Numeracy skills had declined slightly since the last survey in 2003. Seventeen million adults in England in 2011 (just under half the working-age population) were at 'Entry Levels' in numeracy – roughly equivalent to the standards expected in primary school. Further, the survey showed that 78% of the working-age population were at or below level 2 numeracy. These people may not be able to compare products and services for the best buy, or work out a household budget; essentially they would not achieve a good mark in a mathematics examination at age 16.

More generally, the OECD published the results of a survey in October 2013[2] showing that 19% of adults in 21 countries had mathematics skills at the level of a 10-year-old. These adults could only manage one-step tasks with sums, sorting numbers or reading graphs: many could only perform sums with money or whole numbers. For individual countries, the percentage at this level of numeracy ranged from 8% in Japan to 32% in Italy. A selection of the results is shown in Figure 1.1. For England and Northern Ireland, the estimate is that around 8.5 million adults are at this level.

Presenting Data: How to Communicate Your Message Effectively, First Edition. Ed Swires-Hennessy.
© 2014 John Wiley & Sons, Ltd. Published 2014 by John Wiley & Sons, Ltd.

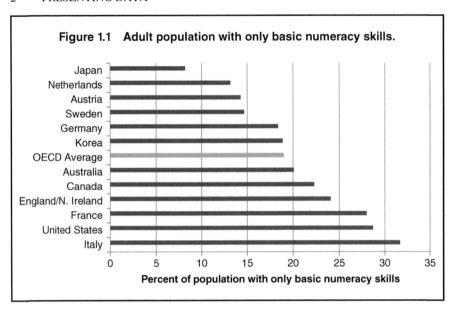

Figure 1.1   Adult population with only basic numeracy skills.

1.1   Thousands separator

When considering a number with more than three digits, most of us will not take in the whole number but will try to understand it. If the number is just a set of digits together without any separators, most will try to find the place for thousands separators to form the basis of understanding the number. Then, depending on the size of the number, will try to do some rounding to understand the number and put it into our mental understanding range.

Example 1.1 ....

Let us consider a number which represents the population of a country.

12573981

How do we try to understand the number? None could honestly say they could read the digits as given and fully understand it without any other process going on in our brains.

So what do we do?

First our brains need a little help. We can introduce separator characters to split the digits into groups of three, starting from the right-hand side. These separators are known as 'thousands separators'. In some countries (United States and United Kingdom), the separator is a comma; for much of Europe, the separator is a space but in some European countries a full stop is used. I visited one country's statistical office website and found three different practices: comma, space and nothing used for the separator! Whichever is chosen, it must be consistently used throughout an organisation's output.

The number, with these separators would then appear as either

$$12{,}573{,}981 \quad \text{or} \quad 12\ 573\ 981 \quad \text{or} \quad 12.573.981$$

This process here helps us to identify that we have 12 million, 573 thousands and 981 units.

But how does our brain handle all this information? It doesn't. The brain will not retain all of the information but introduce some abbreviation of the data. This could either be by truncating the number (i.e. throwing some digits away): for example, after the millions – giving 12 million. Many will argue that a lot of information has been thrown away: but, does it matter? Only, one could argue, if it affects the understanding. Here the number is actually nearer to 13 millions: so would it be better to think of the number in this way? Truncating is easy but may not always be helpful in aiding our understanding. Without another number to compare this one with, it is probably acceptable.

Another approach would be to round the number. Rounding is the process where we reduce the number of digits and say that the number is 'nearly' or 'about' something. The difficulty with rounding is that few know how to do it well!

Should the number be rounded to millions? Hundreds of thousands? Or thousands?

Looking independently, just at this one number, most would round the number to 12.6 million. Is this right? It may be. Let us leave the question for a little while.

Principle 1.1: Choose one symbol for the thousands separator and use consistently.

....

1.2 Decimal separator

Just as differences of symbol exist for thousands separators, some differences exist for the decimal separator. In the United States and the United Kingdom, a full stop is always used as a decimal separator; in Europe, some countries use the full stop and some a comma. Whichever is used, again, it must be consistently used throughout an organisation's output. One example from a national statistical office showed the use of both a comma and a full stop in the same table! An extract of the table is given in Table 1.1. Note also, in this example, that different rounding has been applied to data in lines two to four.

Table 1.1   Different decimal separators used in the same table.

Basic social indicators

Demographic indicators	1970–75	1980–85	1993–99	2005
Life expectancy at birth (years)	66.6	69.7	66.9	70.6
Infant mortality (per thousand live births)	80.8	46.8	47	65,1
Birth rate (per thousand people)	37	40	32.6	31,4
Death rate (per thousand people)	8.1	7	5.3	4,4

Principle 1.2: Choose one symbol for the decimal separator and use consistently.

1.3   Level of detail in comparisons

In seeking to communicate data, it is important to have the same denomination and rounding of data in any data to be compared. So it is more difficult for the user to compare £1.8 billion of turnover with a profit of £232.9 million: it should be phrased as £1.82 billion of turnover and £0.23 billion of profit. Similarly, comparing profit across years, an increase from £246.4 million to £386 million would be better as 'an increase from £246 million to £386 million.' An even worse example of presentation would be to compare the number of homes in the stock, 1.8 million, with a new build figure of 9,768 in the last year: this means nothing to most readers and would be better either comparing the number of this year's newly built homes with that of the previous year or noting a percentage increase (though this would be very small).

In Table 1.1, some of the data is given to one decimal place and some to whole numbers. This may be a result of using Excel which would normally drop trailing zeros unless forced to use the same number of decimal places in the data.

In one press release I found the following statement:

> 11.048 million NHS sight tests, an increase of 563 thousand (5.4%) on the previous year.

Here the user is expected to convert the second number to millions. We have spurious accuracy and difficulty in understanding. This would be better as:

> The number of NHS sight tests given increased by 0.6 million (5%) to 11.0 million.

Principle 1.3: Always provide data for comparisons in the same units.

1.4   Justification of data

When we did our first mathematics lessons in school where we had to add numbers of a magnitude over 100, we were taught that we should start by putting three letters in a row as:

h t u

These letters represent hundreds, tens and units. We were encouraged to write the numbers neatly in the correct columns below the letters. This places the digits of equal value in the same column.

The use of different typefaces in the preparation of tables is easy today. Personal computers have over a hundred preinstalled typefaces which can be used in basic word processing and spreadsheet packages. Choosing the wrong typeface can interfere with our understanding of the presented data: many examples can be seen in simple tables in magazines or newspapers.

Example 1.2 ....

Consider the following data:

$$20,011$$

$$111,111$$

$$88,444$$

Here, using the Constantia typeface, the proportional spacing for the digits has confounded the positioning of the columns of data. Also here the greater height of the '8' draws attention to them. Why is this important? When asked to identify the largest number in a column of data, we have been taught to look to the left of the column of numbers: the one that is further left (if the data are right justified) is, we believe, the largest number.

This, however, relies on a principle of presentation that more and more people are ignoring. In visits to many official government websites, I can find data centred or left-justified in columns of tables (a simple one-button operation in Excel): such presentation is definitely wrong. Let me explain why.

In this example, using this principle, we would conclude that the bottom number is the largest as its first digit is further to the left than the first digit of the first number. Examining the data clearly indicates that this conclusion is wrong as the number in the middle has an extra digit and even that is too far to the right.

....

Principle 1.4: Ensure typeface used for data presentation has the same width for all digits.

> Principle 1.5: Right justify data in columns, ensuring digits of equal value are under similar valued digits in the other numbers.

One other poor presentation of data is often seen in the investment sector. Performance data for various funds are given to differing numbers of decimal places. In two cases recently, not only were the data so presented but, in addition, the data were centred within columns as shown in Example 1.3.

Example 1.3 ....

(a) Centre-justified and different rounding

Fund	Performance over 1 year
A	−46.2%
B	138%
C	125.92%
D	111.365%
E	−31.7%
F	104.25%

If the numbers are right justified, we have:

(b) Right-justified and different rounding

Fund	Performance over 1 year
A	−46.2%
B	138%
C	125.92%
D	111.365%
E	−31.7%
F	104.25%

Now the investor is looking for the largest return. Using the principle of looking to the left of the column, fund D is the highest performance. But looking carefully at all of the data, this is clearly wrong. Here one needs either to round the data to the same level or align the data on the decimal point. Given the variation in the data, the preference here would be rounding to whole numbers as in c. Note the removal of the % symbol from each data item and the change to the column heading.

(c) Right-justified and similarly rounded

Fund	Performance over 1 year (%)
A	−46
B	138
C	126
D	111
E	−32
F	104

. . . .

Principle 1.6: Where possible, round data to be compared to the same level.

1.5   Basic rounding

The main purpose of rounding data is to put the numbers into a memorable and understandable range. Nobody would quote the population of the world to the last digit as it is increasing every second. Also, to write a number with 10 digits is beyond most people's comprehension. According to the United Nations[3], the world's population reached 7 billion on 31 October 2011 and by 22 October 2012, it had reached 7,074,173,556. Now taking Principle 1.6 here, we would have 7.0 billion in 2011 and 7.1 billion in 2012.

Computer programs can handle numbers less than or greater than 0.5 easily: those less than 0.5 are rounded down; those greater than 0.5 are rounded up to the next whole number. However, most computer programs would round any number with an exact 0.5 up to the next whole number. But what is correct? If we start with a line from 0 to 1 and divide it into equal small chunks from either end, we will find that 0.5 is exactly in the middle – neither nearer to the 0 or the 1. So which way should it be rounded? To avoid bias, numbers ending in exact '.5' should be rounded to an even digit.

3 divided by 2 = 1.5 (exactly); rounded to a whole number gives 2
5 divided by 2 = 2.5 (exactly); rounded to a whole number gives 2
7 divided by 2 = 3.5 (exactly); rounded to a whole number gives 4

This principle is easily extended to other numbers: 0 to 4 would be rounded down; 6 to 9 would be rounded up. Similarly, if rounding to the nearest thousand, 0 to 499 would be rounded down and 501 to 999 would be rounded up. So,

14,245 is rounded to 14,000
14,675 is rounded to 15,000

But what happens when we have an exact 500? Technically, it is neither nearer the lower thousand nor the higher one. Again, the principle to be employed should be to round to an even digit. Thus

14,500 is rounded to 14,000

123,500,000 is rounded to 124,000,000 or written as 124 million.

One of the most frequently asked questions with regard to rounding is 'To what level should I round?' In Example 1.1, it was suggested that most would round the number to 1 decimal place of millions. But that is just one number on its own.

Now, instead of just considering one population number, let's consider two.

Example 1.4 ....

12573981 and
111894397

When most people see numbers like this, they ignore them – as they are outside their mental understanding. These numbers also start at the same position on the page but, using Principle 1.5, we need to move the bottom number one digit position to the left.

12573981 and
111894397

Then it is clear that the bottom number is the larger of the two as it has a digit further to the left than the first number.

The first task in trying to understand the data is to add thousands separators. So, introducing these, we have:

12,573,981 and
111,894,397

If we round these numbers as in Example 1.1, we would have:

12.6 million and
111.9 million

We can now ask if we need the digit representing the hundreds of thousands: probably not. These would then round to

13 million and
112 million

....

Such rounding puts the numbers within the boundaries of our understanding and they are much more memorable than the raw numbers. But what is the appropriate level of rounding for numbers? Before answering that, let me give another example.

Example 1.5 ....

The population of Wales in 2012 was estimated to be 3,074,100; the number of sheep in Wales was estimated to be 8,898,200.

Neither of these numbers is particularly memorable as written in the above paragraph. However, if we said:

> The population of Wales is   3 million
>   and
> The number of sheep in Wales is   9 million

These numbers would be in our minds for some time. If we added another statement, we would communicate even better – and make the information totally memorable:

> For every person in Wales, there are 3 sheep!

....

So how should the numbers be rounded to be as helpful to the recipient as possible? When comparing numbers, it is very difficult for the brain to deal with three digits. The recommended level is two digits: in education-speak, this is taking advantage of the number bonds up to 100. Purchasing a chocolate bar for 65 pence and offering a £1 coin to the shopkeeper, most can work out that the change should be 35 pence. This use of decimal-based currency builds up the number bonds to 100 in our memory so that the differencing of two numbers with two digits is relatively easy.

Beyond the two digits of comparison, most would take out a calculator! When such comparisons are demanded through reports, on company profits for example, many people either ignore the numbers or misunderstand them. Hence a communication failure has occurred.

One concept that has been developed to overcome this issue is effective rounding.

1.6   Effective rounding

Using this concept, numbers to be communicated are presented in such a way that the user can quickly appreciate the message and understand the information. Sufficient digits are given so that the magnitude of the numbers is retained but the information inconsequential to the comparison is discarded.

Let us take two numbers representing the profits of a company in successive years:

Year 1: £13,484.2 million   Year 2: £15,857.4 million.

The user has to concentrate very hard to identify the difference. The first reason this is the case is that the numbers are across the page (and in a report are usually separated by text). The second reason is that the numbers are outside the mental range of understanding of most: so, if time is spent on doing the comparison, both will get rounded in the user's mind before the comparison is done: £13.5 million and £15.9 million – from which we can easily judge which is the larger.

So let us make this a little easier and put the second number under the first.

Year 1: £13,484.2 million
Year 2: £15,857.4 million

With this presentation, we start looking at the left-hand side of the numbers and start comparing digits. The second digit identifies that the Year 2 profits are larger. But still, we have to identify how much larger and retain the information.

Is the decimal part of the numbers important? Not really: dropping them does not make a difference to our understanding.

What about the units of millions? The tens of millions? The hundreds of millions?

As discussed, we need to have two digits for a discriminating comparison: obviously, where the digits are the same, no discrimination is possible.

Taking these numbers, we see that the first digit in both is the same: not effective therefore in discriminating between the numbers. The second digits are different and thus effective in discriminating between the numbers: so this digit is the first effective digit in the comparison. Diagrammatically, we have:

Year 1:   £1¦3¦ 4 8 4 . 2 million

Year 2:   £1¦5¦ 8 5 7 . 4 million.
            ☒ ☑

Where ☒ signifies not effective in discrimination and ☑ signifies effective in discrimination.

The principle is that we are considering number differences (in schools, these are termed number bonds) up to 100; so the next digit will give us that. This is the same whether or not the digit is the same value or not. Here we have a 4 and an 8; they are different. That is to say, once a column of discriminating digits is found, the next column is counted as the second effective column regardless of the digits in the column.

Year 1:   £1¦3 ¦ 4¦8 4 . 2 million

Year 2:   £1¦5¦ 8¦5 7 . 4 million.

☒ ☑ ☑

Having found our two effective digits for comparison, we round to that level

Year 1:   £1¦3 ¦ 4¦8 4 . 2 million

Year 2:   £1¦5¦ 8¦5 7 . 4 million

and have:

Year 1: £1 3, 5 0 0 million
Year 2: £1 5, 9 0 0 million.

This gives us a difference to calculate of

$$59 - 35 = 24$$

And, into the context, a difference of

£2,400 million.

Alternatively, the numbers could be presented as:

£13.5 and £15.9 billion
with a difference of £2.4 billion!

Another example:

People in a city on two census dates:

276,321 and 301,948

Putting them in a column and introducing the digit discrimination:

$$2\,7\,6\,,3\,2\,1$$
$$3\,0\,1\,,9\,4\,8$$

So we round to the second digit:

$$2\,7\,6\,,3\,2\,1$$
$$3\,0\,1\,,9\,4\,8$$

Which gives:

$$2\,8\,0,0\,0\,0$$
$$3\,0\,0,0\,0\,0$$

And then anyone can easily see the change of 20,000 and appreciate how much of a change from the first figure (20/280 = 1/14).

Many will argue that a loss of precision has occurred: that is true – but the increase in understanding and communication of what is happening more than outweighs this. And who can say that the numbers in their full form are accurate to the last digit. This is not to say that the data to the last digit is not necessary: such detail is required for calculations with the data. Further, the full detail of the numbers can be given in a reference table and presented in an Annex.

Finally for this section, a very simple example: two scores on examinations were 73 and 83.

Putting the numbers in columns and looking for the effective digits, we have:

$$7\,3 \text{ and}$$
$$8\,3$$

So rounding to the second effective digit

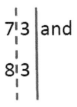

gives

73 and

83

That is, the numbers are left without rounding. Note that here the second digit is the same in each of the numbers. The rule is that once the first column of digits effective for discriminating between the numbers is found, the next column is automatically effective even if it is the same digit.

So, using effective rounding assists the user to understand what message you are trying to get across. It is not a universal panacea to communicating numbers but it certainly helps the provider to consider whether the data are presented in a way they would be understood. Using effective rounding may suggest other issues with the data and may make one wonder whether the data should be included in the presentation.

Let's consider the fairly basic table of population data from 1995 in Table 1.2. The countries included were the countries from which lecturers came to teach on a course in London in 1996.

Table 1.2   Populations of countries, 1996 – 1.

Country	Total population
France	58020376
Denmark	5215732
Germany	81538628
United Kingdom	58491643
Luxembourg	406589
Total	203672968

Now, from the earlier Principles in this chapter, we see that this table has a few problems: the numbers are left-justified; no thousands separators are shown; no rounding of the data has been done.

If we take the issues in order, first let us right-justify the data which gives us Table 1.3.

Table 1.3   Populations of countries, 1996 – 2.

Country	Total population
France	58020376
Denmark	5215732
Germany	81538628
United Kingdom	58491643
Luxembourg	406589
Total	203672968

At least this gives us some sense of the size of the numbers but too much detail is given to really understand the data. So, rounding effectively, we have Table 1.4.

Table 1.4   Populations of countries, 1996 – 3.

Country	Total population (millions)
France	58
Denmark	5
Germany	82
United Kingdom	58
Luxembourg	0
Total	204

The question one would have to ask here is whether it was important to show the data for Luxembourg. In order not to cause an international incident, the answer in this case was that it is important. But still we have a problem: to show Luxembourg as zero is a little of an insult. To overcome this, it is necessary to show the data to 1 decimal place – as in Table 1.5.

As an aside here, it is obvious that the total shown does not equal the addition of the numbers in the table as it is independently rounded from the raw data total. Most people would now understand this but if queries arise, add a footnote to the table indicating that 'The total shown may not be the sum of the constituent parts because of rounding'.

Table 1.5   Populations of countries, 1996 – 4.

Country	Total population (millions)
France	58.0
Denmark	5.2
Germany	81.5
United Kingdom	58.5
Luxembourg	0.4
Total	203.7

Principle 1.7: Use effective rounding of data where possible to aid communication.

Even this table can be improved by sorting the data by either by country or, better in this case, by size of population. This would give Table 1.6.

Table 1.6   Populations of countries, 1996 – 5.

Country	Total population (millions)
Germany	81.5
United Kingdom	58.5
France	58.0
Denmark	5.2
Luxembourg	0.4
Total	203.7

It is then clear that Germany had the highest population and that of the United Kingdom is just higher than that of France. So a message can be more clearly seen in the rounded and sorted table than in the original Table 1.2.

Principle 1.8: In summary tables, consider sorting the data to elicit a message.

Summary of principles: Understanding number

Principle 1.1: Choose one symbol for the thousands separator and use consistently.

Principle 1.2: Choose one symbol for the decimal separator and use consistently.

Principle 1.3: Always provide data for comparisons in the same units.

Principle 1.4: Ensure typeface used for data presentation has the same width for all digits.

Principle 1.5: Right justify data in columns, ensuring digits of equal value are under similar valued digits in the other numbers.

Principle 1.6: Where possible, round data to be compared to the same level.

Principle 1.7: Use effective rounding of data where possible to aid communication.

Principle 1.8: In summary tables, consider sorting the data to elicit a message.

Notes

1. UK's BIS Skills for Life survey 2011. BIS Research Paper 37, December 2011, URN 11/1367 http://www.nationalnumeracy.org.uk/news/16/index.html.

2. Table A2.5, page 242 of http://skills.oecd.org/documents/OECD_Skills_Outlook_2013.pdf.

3. 2010 Revision of the World Population Prospects, UN, http://esa.un.org/unpd/wpp/index.htm.

2

Tables

Before considering issues surrounding the design of tables, remember a couple of other important principles which relate to how the users seek to understand the content of tables.

2.1 Position of totals in tables

The first is the position of totals in a table. If we consider the education of our users, we know a lot about the principles they were taught in school and how they take in information.

Let us start with a set of numbers that is to be added to a total as presented in a row:

$$23 + 16 + 42 + 30$$

If we did a poll in the street and asked people where they would put an equal sign and the total, 100 per cent would say on the right of the row of numbers. Why? Because that is what they were taught in school.

Similarly, if we had a set of numbers in a column as:

$$23+$$
$$16+$$
$$42+$$
$$30$$

Again, if we asked people in the street where they would put the total of the numbers, 100 per cent would say a line should be drawn beneath the column of numbers and the total put below the line.

Presenting Data: How to Communicate Your Message Effectively, First Edition. Ed Swires-Hennessy.
© 2014 John Wiley & Sons, Ltd. Published 2014 by John Wiley & Sons, Ltd.

Most available statistical software presents totals to the left of rows and at the top of columns: why? Simply, I believe, because it is easier to program that way! Others have argued that such presentation puts the important information first but that ignores the fact that most would not look there for totals. Where totals are broken down into constituent parts, some say that the total should be first followed by the 'Of which', as in Table 2.1.

Table 2.1   Distance of travel to work by car.

All distances	5,467
Of which:	
Less than 1 mile	1,567
1 but less than 2 miles	2,358
2 but less than 5 miles	1,021
5 but less than 10 miles	476
10 miles and over	45

But what is the difference that makes this sensible? The table would be easier for most to assimilate the information and understand with the total at the bottom as in Table 2.2.

Table 2.2   Distance of travel to work by car.

	Number
Distance to work:	
Less than 1 mile	1,567
1 but less than 2 miles	2,358
2 but less than 5 miles	1,021
5 but less than 10 miles	476
10 miles and over	45
All distances	5,467

These are not good enough reasons to change our presentation of data to the general user who expects the totals to be to the right of rows and the bottom of columns.

So, in presenting tables with totals to general users, put the totals where users expect them to be – on the right of rows and at the bottom of columns. Going against this principle will confuse users and make them take extra time to understand what is being presented.

Principle 2.1: Put totals to right of rows and bottom of columns of data.

2.2   What is a table?

Many believe that a table is just a collection of data and is a means of organising the data for presentation. Actually two basic types of table can be defined: reference table portraying a great deal of data to the highest precision possible and summary table which is an extract of data from a reference table with some additional work having been done on the information. This additional work could be either rounding of the basic data, sorting of the data, provision of contextual data or simple additional information like percentages.

2.3   Reference tables

Reference tables are just what the adjective implies: tables are there only for reference. Nobody would expect to reproduce a reference table from someone else in their report – they are produced to give a historical record, to set out the maximum possible data set from which data can be extracted to allow the data to be used in further calculations. This last use implies that the data shown in such tables have many non-zero digits, indeed as many as the database has or as many as can reasonably be shown on paper or a computer screen.

The presentation of reference tables is always in the greatest detail and in some form of classification order. So the data for each of the states of the United States in a reference table would normally be presented in an alphabetically sorted list. But, this type of table is rarely useful in communicating a message. The standard order allows both the identification of one number, for example, the population of an area in a certain year, and the easy extraction of data into summary tables.

The reference tables are useful as a means to provide the raw data that are then worked on, rounded, summarised, ordered and often have supplementary information provided alongside. So, in brief, the key elements of a reference table are:

(a) The numbers within the table have as much precision as possible. For example, if one looks at the population of a country, the numbers will be given to the individual person such as – the population of Wales in 2011 was 3,063,456. The rest of the table will include the numbers of people in each county and possibly analysed by sex but all given to the last digit;

(b) The organisation of the numbers in the table is in a classification order, alphabetic or numeric code order;

(c) The number of numbers in the table is not limited; and

(d) The table does not contain any derived information, for example, percentage changes.

These tables are essential to users of data because they allow access to the greatest detail possible for the derivation of summary tables with simple analysis or for use in, say, economic models.

Well-presented reference tables are easy to work with and helpful to those requiring data. Many such reference tables are now available on the websites of organisations and national statistical offices: some are exact replicas of the previous paper versions of tables and some have been adapted to use the facilities available on the web to allow the user to extract just what is required, displaying the data on screen or downloading the data into a spreadsheet. Facilitating the downloading of the information allows the user to format the table in a standard style, either for organisational or personal use.

Some tables that used to be available in paper format are now only available on the Internet in PDF format. Very few such deliverers of information think through the delivery process to the user. Have you ever tried to download a table from a PDF file on the Internet into Excel to work on? Each line of the table is put into one cell[1]!

The first principle of reference table design is consistency. Thus, for every table produced by the organisation, a user only has to learn and understand how they are presented once.

Examples of such tables can be found in the statistical yearbook of a country. These contain many tables, lots of detail and little commentary or explanation. Most of the tables will have data listed according to a classification order, such as the International Classification of Diseases, the Classification of Economic Activities, states or regions. No attempt will have been made to order the data so that a message could be easily seen or derived. That is for the extractor or user of such data.

Even with such tables, some principles of design and consistency will help the user. These are illustrated and then summarised below.

Table number: Table title (a)

Table descriptor (e.g. millions)

Row descriptor	Column heading 1	Column heading 2	Column heading 3
Row heading 1			
Row heading 2 (b)			
Row heading 3			
Row heading 4			
Row heading 5			
Row heading 6			
Total			

Source: Local Government Data Unit ~ Wales

(a) Footnote 1.
(b) Footnote 2.

- Titles should be brief noting the main axes of the table and a period.
- Column headings should be centred over the column.

- Where periods are shown in the table, it is not necessary to repeat them in the table's title. For example, if a table contains population data, by sex and age, for the years 2001 to 2009, the title should be:

  Population by age and sex

  and not

  Population by age and sex, 2001 to 2009.

- Footnote markers are best as lower case letters in brackets and should be put against the column or row heading not in the data cell (because this affects the comparison of data in the table). Try to minimise the number of footnotes, especially against the table title. Their ordering should be: against title; against column headings (left to right) and then against row headings (top to bottom): such ordering allows the user to find the footnote markers easier.

- Where a table of general data has more than five rows, add a space line after the fifth line to break the table in suitable chunks for extracting individual data. It is easy to identify figures in blocks of five without using a ruler. If the data are monthly, it would be more appropriate to add a space line after three rows. Note that, in spreadsheet packages, it is not the insertion of an extra row but making the height of the sixth (or fourth) row twice the height of the normal rows: this avoids extraneous entries appearing in the tables, for example, if calculating a percentage for all rows.

- The table descriptor should only be shown if all of the data in the table are in the same units; otherwise add a descriptor (in parentheses) to each column or row heading as appropriate.

- Source: where data in a report or publication are from a multiplicity of sources each table should have its own source as indicated above. Where all of the data is from the publishing organisation, this is not necessary – but is helpful to users who extract individual tables into their reports since the source is already shown.

- When giving a total's breakdown within the rows of a table, use indents to show the relationship to the heading and put the total at the bottom of the section. For example:

  Mode of travel to work:

      Car as driver

      Car as passenger

      Bus

      Train

      Bicycle

      Other

      All modes

Principle 2.2: Design reference tables with some understanding of expected use.

2.4   Summary tables

The basic rationale for producing summary tables is to present information with a message that can be easily assimilated by the audience. If the audience is a group of senior people in an organisation with little time to digest the message from data, it is the producer's responsibility to spend time reducing the reference table into something that can be quickly and easily understood. Some managers insist on having lots of detail in tables: but to what end? Can they readily understand all of the data? I doubt it – but it does give some 'comfort' that they can always look up data: so give them the detail in an Annex or a computer file and present them with summary information. In the Internet age, the ultimate would be a displayed summary table with links to the detailed data.

Many people faced with a reference table or a large summary table that fills a page will quickly turn the page rather than try to understand what message the data show.

A small summary table, however, with a specific purpose can often drive a message home. Identifying that purpose before starting to construct the table will help its design. The important questions are:

(a) Which numbers do I want the reader of my table to compare?

(b) How many numbers are necessary to meet the aim of the table?

(c) How much detail should each number have?

(d) Can I add some derived information to help the identification of the message?

(e) Should the numbers be sorted in some way?

These questions – essentially criteria – should be the basis for all summary tables. Considering these in the design of the summary table will assist in achieving the best possible table for the purpose given.

Example 2.1 ....

Let us look at the Gross Value Added per head information in Table 2.3 which is presented in the standard classification order of regions/countries for the UK.

Table 2.3   Gross Value Added per head.

£

	2006	2007	2008	2009	2010
North East	14,901	15,530	15,673	15,304	15,744
North West	16,382	17,165	17,344	16,884	17,381
Yorkshire and The Humber	16,227	16,900	17,012	16,512	16,917
East Midlands	17,013	17,806	17,952	17,519	18,090
West Midlands	16,365	17,098	17,143	16,602	17,060
East of England	18,514	19,337	19,294	18,536	18,996
London	31,714	33,721	34,964	34,779	35,026
South East	20,472	21,593	21,859	21,257	21,924
South West	17,576	18,383	18,606	18,184	18,669
Wales	14,407	15,042	15,122	14,664	15,145
Scotland	18,484	19,492	19,991	19,755	20,220
Northern Ireland	15,359	16,013	15,928	15,249	15,651
United Kingdom	19,542	20,539	20,911	20,341	20,849

This is an extract from the reference table which showed 11 years of information. What is the aim of the summary table to be prepared? To highlight the differences in gross value added per head across the regions/countries of the United Kingdom. So, looking at our criteria for a summary table, we have:

(a) The numbers are in a column – so easy to compare;

(b) Five numbers per area are given – so some redundancy: not all numbers for each area are required to tell the story;

(c) The values given are to the nearest one pound: this is not necessary. Applying effective rounding, we only need the thousands, that is, the first two digits of each rounded number;

(d) No derived information is given: but we could add a column indicating the relative size compared with the United Kingdom average;

(e) And we could sort the table, from the highest to the lowest.

Applying these answers to Table 2.3 produces a very different table – Table 2.4.

Table 2.4   Gross Value Added per head, 2010.

	£'000	Per cent of UK average
London	35	168
South East	22	105
Scotland	20	97
East of England	19	91
South West	19	90
East Midlands	18	87
North West	17	83
West Midlands	17	82
Yorkshire and The Humber	17	81
North East	16	76
Northern Ireland	16	75
Wales	15	73
United Kingdom	21	100

From this resulting table, it is clear that London and the South East are the only regions with Gross Value Added per head above the UK average – London significantly – and that the area with the lowest value is Wales. The information from this table is easier to understand, clearer and quicker to derive the message from.

Some would argue that the loss of the precision and the absence of at least one decimal place in the percentage column is a step too far. However, the full numbers are not lost and, if necessary, could be put into an annex of the document. With regard to the percentage rounding, one can only ask if the table meets the purpose: with the sorting, it does, so additional detail is unnecessary.

. . . .

Principle 2.3: Design the summary table to meet a specified purpose – including observance of the principles noted in Chapter 1.

2.5   How tables are read

A second principle to take into account for tables is how the user will read the table. When faced with a table, we do not go to the middle of a table and try to find a figure. It is read as any other text – from left to right and from top to bottom. So, in terms of information flow, this should be taken into account. Here, a time component is an example of where this is true. When time is a component of a table, it should always go from left to right or from top to bottom of the table, that is, the latest data should always be on the right of a row or the bottom of a column. Which of these is better? It really depends on what you are asking the user to do with the data: if you want them to compare one series of data over time, it is better to have time running down a column. If the main purpose of the table is to show what elements constitute a total, then time would go along a row.

> Principle 2.4: Present the table with time running left to right or top to bottom.

2.6   Layout of data in tables

For summary tables, it is exceptionally important to understand the purpose of the table as this will influence the layout. Let us consider the basic information in Table 2.5.

Table 2.5   Motor vehicles currently licensed in UK.

Thousands

Type of vehicle	1961	1966	1971	1976	1981
Private cars and private vans	6,114	9,747	12,361	14,373	15,632
Motorcycles, scooters and mopeds	1,842	1,430	1,033	1,235	1,386
Public transport vehicles	94	96	108	115	112
Goods	1,490	1,611	1,660	1,796	1,771
Agricultural tractors, etc.	481	478	450	414	373
Other vehicles	206	260	247	300	510
All vehicles	10,227	13,621	15,859	18,233	19,784

Source: Plain Figures

Here time runs correctly across the page, earliest to latest. We are asking the user to understand what is happening to licensed traffic as a means of understanding why roads are becoming more crowded. So we want the user to compare numbers across the rows: 6,114 with 15,632. Actually, this is quite difficult for most. Let us round the data effectively (see Section 1.6) and turn the table round so that time runs down the page: this gives Table 2.6.

Table 2.6   Motor vehicles currently licensed in UK.

Thousands

	Type of vehicle						
Year:	Private cars and vans	Motorcycles, scooters, mopeds	Public transport	Goods vehicles	Agric. tractors etc	Other vehicles	All vehicles
1961	6,100	1,840	94	1,490	481	206	10,200
1966	9,700	1,430	96	1,610	478	260	13,600
1971	12,400	1,030	108	1,660	450	250	15,900
1976	14,400	1,240	115	1,800	414	300	18,200
1981	15,600	1,380	112	1,771	370	510	19,800

Source: Plain Figures

Now it is quite easy to see what is happening: the principle here is that putting data to compare in columns is much easier for the user to assimilate the information than if the data were in rows. Even this version of the table presents the user with a lot of information which is actually unnecessary if the purpose of the table is to understand why roads are becoming more crowded. Suppose the table was for presentation to a member of the government: very little effort would produce the simplest table as Table 2.7.

Table 2.7   Motor vehicles currently licensed in the UK.

Millions

	Private cars and vans	Other vehicles
Year:		
1961	6	4
1981	16	4

Here it becomes very clear that the increase in traffic was due to the more than two and a half times increase in the number of private cars and vans since the number of 'Other vehicles' stayed a similar number. The bullet point summary of this table would then be:

- Car and van numbers more than doubled;
- Other vehicle numbers essentially static.

When writing the description of the data as in these bullet points, the producer has a pointer as to which data are important in the summary table. Should the initial summary table contain more data than is referred to in the bullet points, these can be removed from the table and Table 2.7 would be the result.

The most powerful table contains only two numbers in a reasonable amount of white space on a page. Obviously, the numbers have to be presented properly taking in the principles of Chapter 1. Look again at Table 2.7. Should the second column of data be removed and the table prefaced by a statement as to what the data are, this would produce . . .

The numbers of cars and vans licensed in the United Kingdom grew significantly from 1961 to 1981 as:

	Millions
1961	6
1981	16

This presentation concentrates the mind of the user on the most significant aspects of the change in vehicles licensed.

Producers often make life difficult for users of tables by not presenting them in the clearest way. The next two tables ask the user to compare numbers which are not proximate: the first is asking users to compare alternate numbers in columns, the second in rows.

Table 2.8 mixes numbers and rates for a particular disease. It is meant to be a summary table but the inclusion of both numbers and rates in the same cell is making the message unclear.

Table 2.8   A disease in males.

Number and rate

	South	North	Mid & West	Total
15–44	1263	610	924	2797
	5.67	4.5	4.85	5.1
45–54	1744	1022	1492	4288
	7.97	7.54	7.83	7.81
55–64	4458	2430	3553	10441
	20.02	17.93	18.64	19.03
65–74	7987	4819	6965	19771
	35.87	35.56	36.54	36.03
75–84	5558	3757	5106	14421
	24.96	27.73	26.79	26.28
85–99	1228	912	1019	3159
	5.51	6.73	5.35	5.76
Total	22268	13550	19059	54877
	100	100	100	100

In Table 2.8, several issues arise.

- Alternate rows are different types of data and thus hinder the direct comparison in the column; the rates should be separated out to a new table;

- The numbers in the table do not have thousands separators and are centred in the columns;

- Some data are rounded differently (this was most likely prepared in Excel and the originator has not forced the rates data to have the same number of decimal points);

- The gridlines in the table are disruptive to the eye when comparing across cells;

- Whenever a total of percentages is shown, it should be to the same number of decimal places as the component data that is, in this case, 100.00. In this table, one decimal place is sufficient for the differentiation across areas. However, if the purpose of the table was to show the significant differences across ages, no decimal places would be required.

The second example is where the originator of a table tries to be helpful by adding other information to show some characteristics, for example, percentages. Consider Table 2.9.

Table 2.9   Example of percentage data interfering in message.

	1991		2001		2011	
Distance to work:	Number	Per cent	Number	Per cent	Number	Per cent
Less than 1 mile	1,567	28.7	2,687	30.7	4,324	33.1
1 but less than 2 miles	2,358	43.1	2,937	33.6	3,567	27.3
2 but less than 5 miles	1,021	18.7	1,854	21.2	2,875	22.0
5 but less than 10 miles	476	8.7	843	9.6	1,358	10.4
10 miles and over	45	0.8	432	4.9	943	7.2
All distances	5,467	100.0	8,753	100.0	13,067	100.0

In this case, again, the user is expected to jump over either the percentage data to compare numbers or jump over the numbers to compare percentages. Here, inverting the table so that time runs down the page and splitting the numbers and percentage data would give Table 2.10 which is easier to read and understand.

Table 2.10   Distance of travel to work by car.

	Less than 1 mile	1 but less than 2 miles	2 but less than 5 miles	5 but less than 10 miles	10 miles and over	All distances
Numbers:						
1991	1,567	2,358	1,021	476	45	5,467
2001	2,687	2,937	1,854	843	432	8,753
2011	4,324	3,567	2,875	1,358	943	13,067
Percentages:						
1991	28.7	43.1	18.7	8.7	0.8	100.0
2001	30.7	33.6	21.2	9.6	4.9	100.0
2011	33.1	27.3	22.0	10.4	7.2	100.0

In this presentation of the data, the percentages are together and it is easier to identify trends.

Too often tables are thrown together without much thought. How will they fit on the page? Are the headings too big for columns? Do the data fit better one way round? And yet few build the real purpose of the table into the design.

Thus, in designing a table it is necessary to have some regard for the purpose of the table, how the user will read it and seek to understand it. Thinking these ideas through will influence the design of tables and keep the user with the purpose of the table and continuing to read.

> Principle 2.5: Put into columns the categories of the data that you want the user to compare.

2.7   Capital letters for table titles and headings in tables

Some organisations use capital letters throughout words in the titles of tables and in the column headings. It is believed that this makes the words stand out more. However, for the user, this is making the task of reading and understanding significantly harder. Miles Tinker, renowned for his landmark work, *Legibility of Print*[2], performed scientific studies on the legibility and readability of all-capital print. His findings were as follows:

> 'All-capital print greatly retards speed of reading in comparison with lower-case type. Also, most readers judge all capitals to be less legible. Faster reading of the lower-case print is due to the characteristic word

forms furnished by this type. This permits reading by word units, while all capitals tend to be read letter by letter. Furthermore, since all-capital printing takes at least one-third more space than lower case, more fixation pauses are required for reading the same amount of material. The use of all capitals should be dispensed with in every printing situation.'

Colin Wheildon[3] in later studies concluded that headlines set in capital letters are significantly less legible than those set in lower case. Basically, when reading text, we do not read every letter individually but recognise a word 'pattern'. This is fine for words in lower case type as we see and recognise so many. However, we are not used to reading words in capital letters and thus we have to look more to each letter. A study in 1928 showed that 'all-capital text was read 11.8 per cent slower than lower case, or approximately 38 words per minute slower,'[4] which means we take an appreciably longer time to take in such words.

> Principle 2.6: Don't use capital letters for table titles or headings in tables.

2.8   Use of bold typeface

Another common misconception is that the use of a lot of bold typeface in tables is good and draws attention to specific figures. This is often to emphasise some part of the table, for example, the total line including both text and numbers. If the emboldening of the numbers moves them from the correct alignment within columns and the producer is asking the user to compare numbers in the columns, it is preferable not to embolden the numbers. Also, a table with too much of the text emboldened negates the purpose of emboldening and makes the prime message more difficult to capture.

> Principle 2.7: Use emboldening sparingly, if at all.

2.9   Use of gridlines and other lines in tables

Gridlines in tables can be regarded as clutter and something to be avoided. Use of them essentially puts a barrier between each number the user is trying to compare and does not help. The use of appropriate space lines in tables to break tables into meaningful sections or to facilitate the extraction of numbers from large tables is preferable.

The use of a line in a table to break up sections should be used only when a change of definition needs to be indicated – which would be footnoted.

Again, in large tables, some have introduced shading of alternate lines in different colours to assist with the reading of the table. Whilst for some tables this does make reading them a little easier, more thought in the table's construction and the use of space lines would do the same job without costing more time in setting the shading on the rows. It is also important, if a colour is used as shading, to photocopy a page or print one on a black and white printer to see the effect. Obviously, if the data are obscured, some adjustment of the colour needs to be effected.

> Principle 2.8: Don't use gridlines in tables and use other lines sparingly.

> Principle 2.9: If using a colour as background to rows or columns in a table, check that the table will print clearly on a black and white printer.

Summary of principles: Tables

Principle 2.1: Put totals to right of rows and bottom of columns of data
Principle 2.2: Design reference tables with some understanding of expected use.
Principle 2.3: Design the summary table to meet a specified purpose – including observance of the principles noted in Chapter 1.
Principle 2.4: Present the table with time running left to right or top to bottom
Principle 2.5: Put into columns the categories of the data that you want the user to compare.
Principle 2.6: Don't use capital letters for table titles or headings in tables.
Principle 2.7: Use emboldening sparingly, if at all.
Principle 2.8: Don't use gridlines in tables and use other lines sparingly.
Principle 2.9: If using a colour as background to rows or columns in a table, check that the table will print clearly on a black and white printer.

Notes

1. The only exceptions to this are if the user has a full copy of Adobe Acrobat which allows full access to all elements of documents – thus enabling the extraction of a table in proper format – or a special program. Neither of these two options is common for general users.

2. Tinker, M.A. (1963). *Legibility of Print*. Ames, Iowa: Iowa State University Press. p. 65. ASIN B000I52NNE

3. Wheildon, C. (1995). *Type and Layout: How Typography and Design Can Get your Message Across - Or Get in the Way*. Berkeley, CA: Strathmoor Press. p. 74. ISBN 0-9624891-5-8.

4. Tinker, M.A. and Paterson, D.G. Influence of type form on speed of reading. 1928, *Journal of Applied Psychology*, 12, 359–368, in Miles Tinker, *Bases for Effective Reading*, 1965, Minneapolis, Lund Press. p. 136.

3

Charts (bar charts, histograms, pie charts, graphs)

Many charts presenting data display a wrong or inappropriate message. This can be either because of the poor presentation via scale differences or truncation or because the data are not represented well. This chapter will look at the rationale for charts, show through examples why certain charts are not good and illustrate how we can simply improve some charts with a little thought. User perception plays a large part in understanding charts, and illustrations of how the wrong message can be taken from charts are shown.

3.1 How the user interprets charts

When a user is shown a chart, the most usual approach is to seek to understand the visual display of the data without recourse to the axes, footnotes or other help that may have been available. So, if the chart does not represent the data well, the wrong message may be assimilated. We often make the presumption that the originator of a chart will understand the data and present the data within a chart in the way it should be interpreted. However, experience tells me that not only do newspapers present poorly designed and constructed charts but so do National Statistical Institutes.

The basic presumption of this chapter is that the chart's originator is trying to represent the data well and show a message from doing so. Otherwise, one has to ask why the chart has been included in a document. From some challenges I have put to originators, the answers as to why some awful – but very colourful – diagrams have

Presenting Data: How to Communicate Your Message Effectively, First Edition. Ed Swires-Hennessy.
© 2014 John Wiley & Sons, Ltd. Published 2014 by John Wiley & Sons, Ltd.

been constructed and issued have been as simple as:

- They enliven the text.

- The colour brings the report to life.

- The charts look modern and interesting.

Further, the choice of three-dimensional charts is often because someone has suggested that such a presentation will make the chart interesting or enhance the document in which it is placed. None of these reasons are sufficient to justify the inclusion of the poor charts and the basic rationale will be explained in this chapter.

Let us start with a basic pictogram example in Figure 3.1. We are trying to illustrate the difference between two bonus payments in the form of a pictogram.

Figure 3.1   Bonus bags.

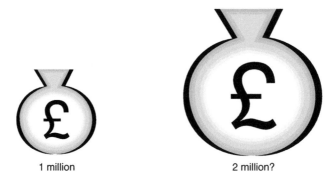

The bag on the right is twice the size of the first in width terms. But is this the right representation? No, because we are trying to represent a single variable in an area which is a two-dimensional space.

Looking at the second of the money bags, we have the following explanation.

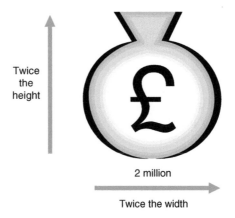

The area of the bag, relative to the first, is therefore $2 \times 2 \times$ first bag $= 4$ times the area of the first bag! This does not represent the data well.

What should be presented here is a picture of two bags, one for each of the two million as in Figure 3.2.

Figure 3.2   Proper pictogram display.

1 million   2 million

We shall come back to pictograms later in the chapter.

An old saying is

A picture is worth a thousand words.[1]

For us, in trying to communicate data, I would change the phrase to

A picture is worth a thousand words if and only if it is the correct picture.

By 'correct' here, we mean that the picture accurately and clearly represents the data and the message from them. Throughout this chapter, I shall apply the phrase to bar charts, graphs and pictograms.

Principle 3.1: A chart must represent the data clearly and accurately.

3.2   Written aims for charts

When the majority of people are asked to create a chart representing data, the first action is to start up the chart-making program. This should not be the first step. Instead, words should be written down explaining the aim in very specific terms for the chart: there is no substitute for this step.

Once the written aim is available, the creation of a chart can begin. Once completed, the finished chart can be examined in the light of the aim to see if it meets the aim. If it does, work can begin on the next chart. If it doesn't, then the chart must be scrapped or changed so that the aim is met.

Another good reason exists for writing aims. If one person was responsible for producing a regular booklet or bulletin with data and charts, he or she should understand why the tables and charts were produced in the presented format and design. If that person moves on to another job, the history will be lost and the easiest route to creating the next edition will be to delete one year's data and add the latest! Without a written aim to measure the revised charts against, how can we know they are purposeful and correct?

Let's look at an example of travel to work data with a written aim of:

To show the differences in modes of travel to work **within** counties

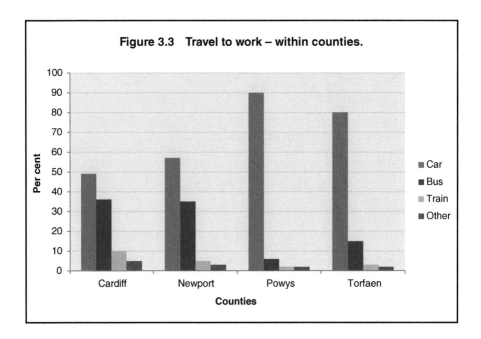

Here, the clustering of the data is mode within county, with the legend showing the different modes.

If we now change the aim for the chart to:

To show the differences in modes of travel to work **between** counties

We are then interested to see how the modes differ between the counties and thus the clustering is by mode not by county as in Figure 3.3. This is shown in Figure 3.4.

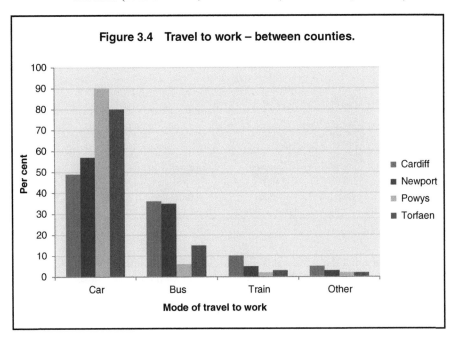

Figure 3.4   Travel to work – between counties.

Comparing Figures 3.3 and 3.4, it is clear that they are very different and have different messages for the user. If the wrong one is presented, the user may be misinformed or misled. More on this topic will be discussed throughout this chapter.

Principle 3.2: Before starting to create a chart, have a clear and specific written aim.

3.3   Scale definition and display

When the user is presented with a chart, it is common for most to spend little time trying to understand it. The user seeks to understand the 'picture' in the chart without recourse to the scales presented or the notes that may appear underneath the chart. So, let's consider Figure 3.5.

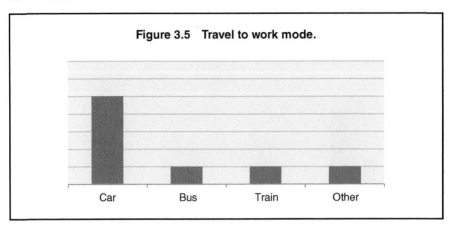

Here we deduce that the car mode is five times greater than bus or train or other. If we present a chart as in Figure 3.5, the visual deduction of the ratio of those travelling to work by car to those travelling by bus is 5 to 1. If we now add the scale and read the values of the bars from the scale in Figure 3.6, we get a different message: that the percentage travelling to work by car is twice the percentage travelling by bus, that is, a ratio of 2 to 1.

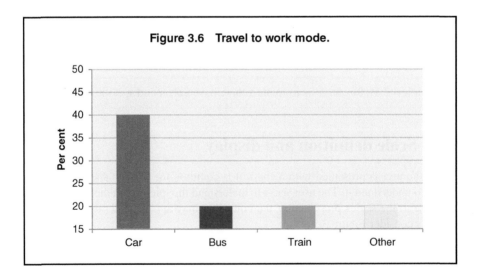

And the proper presentation of the information in a chart is given in Figure 3.7, where the y-axis scale starts at zero and allows a proper visual comparison of the sizes of the bars.

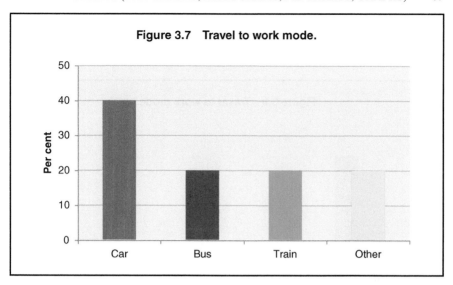

Figure 3.7   Travel to work mode.

In Figure 3.7, each of the bars is separate and thus the area of the bar, relative to the other bars, is clear. If we add an extra set of data for a later year, it is easy to overlap the bars as in Figure 3.8.

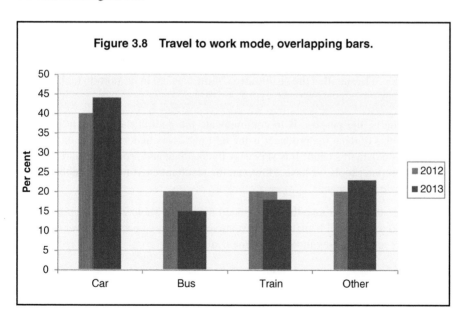

Figure 3.8   Travel to work mode, overlapping bars.

This, however, actually means that the interpretation of the chart is made more difficult. We are asking the user, for example, to judge the relative sizes of the two bars with one partly hidden. So, instead of comparing the areas of the bars, the user can realistically only compare the heights of the bars. Depending on the colouring used, one set of bars may be more dominant to the eye through colouring. The proper presentation of the 2 years of data is given in Figure 3.9.

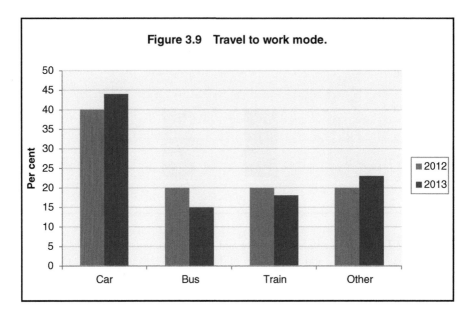

Figure 3.9   Travel to work mode.

Principle 3.3: Ensure that the visible areas of the bars are in proportion to the data.

With all charts, the purpose is to communicate the data in the best way possible. As with tables of data, the producer is the best person to understand the data and to present those data in a chart. It is very easy to take data and produce a chart with the modern computer tools we have. However, the resultant chart may not be the best to communicate a message.

To illustrate this, let us take the value added data from Table 2.3 in Chapter 2. The aim of such a chart will be to show the relative positions of each area over time and identify the areas needing more government support.

Table 2.3   Gross Value Added per head.

£

	2006	2007	2008	2009	2010
North East	14,901	15,530	15,673	15,304	15,744
North West	16,382	17,165	17,344	16,884	17,381
Yorkshire and The Humber	16,227	16,900	17,012	16,512	16,917
East Midlands	17,013	17,806	17,952	17,519	18,090
West Midlands	16,365	17,098	17,143	16,602	17,060
East of England	18,514	19,337	19,294	18,536	18,996
London	31,714	33,721	34,964	34,779	35,026
South East	20,472	21,593	21,859	21,257	21,924
South West	17,576	18,383	18,606	18,184	18,669
Wales	14,407	15,042	15,122	14,664	15,145
Scotland	18,484	19,492	19,991	19,755	20,220
Northern Ireland	15,359	16,013	15,928	15,249	15,651
United Kingdom	19,542	20,539	20,911	20,341	20,849

Simply charting this data produces the rather nonsensical Figure 3.10. Here, most of the lines are close together making it difficult to identify which is which and the default ordering of the colours implies, at first glance, that the top line is that of North East whereas it is actually that of London. The message from the chart is that most of the areas are similar over time with London significantly higher than the rest. Further, with this dataset, it is very difficult to identify clearly all of the separate colours against their headings; thus, the aim of indicating the relative position of the areas is not achieved.

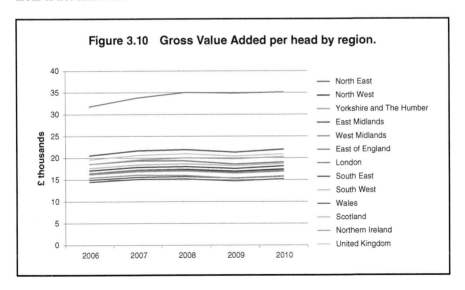

Figure 3.10   Gross Value Added per head by region.

Examining this chart closely shows that the relative position of each area appears to change only marginally over time. So let's change the aim to looking only at the last year of data and examining each area relative to the overall average for the United Kingdom. The data table is shown in Table 3.1.

Table 3.1   Relative Gross Value Added per head, 2010.

	United Kingdom = 100
North East	76
North West	83
Yorkshire and The Humber	81
East Midlands	87
West Midlands	82
East of England	91
London	168
South East	105
South West	90
Wales	73
Scotland	97
Northern Ireland	75
United Kingdom	100

From these data, we can produce Figure 3.11. This meets the aim we set but could it be better?

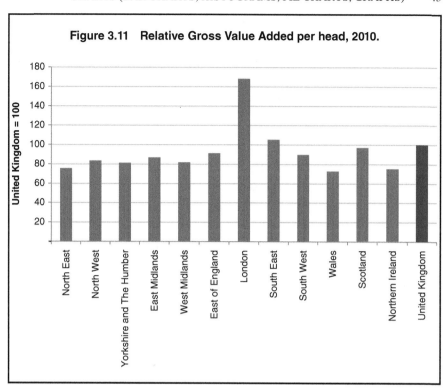

Figure 3.11   Relative Gross Value Added per head, 2010.

Look back to what was done in Chapter 2. The data for 2010 in Table 2.4 were not left in the same order as in Table 2.3: they were sorted, largest to smallest. Some would argue that the order of data in Table 2.3 is the most appropriate as it is in the standard presentation order for such data. However, the message from the data is much more difficult to identify. Sorting into value order enables the user to identify clearly whether West Midlands had a higher Gross Value Added per head than the North West.

Principle 3.4: Arrange the elements of the chart to maximise understanding of the information.

Another issue here is the reading of the text: quite difficult in this position. So let us turn the diagram through 90 degrees so that the text is horizontal. And, finally, to emphasise the United Kingdom as an overall average, instead of having a bar for the country, we could add a line to indicate clearly the average. Hence, we can quickly see those above average and those below. This is shown in Figure 3.12. So the sorting here not only helps the user to understand the message but also assists the producer to derive the message from the data.

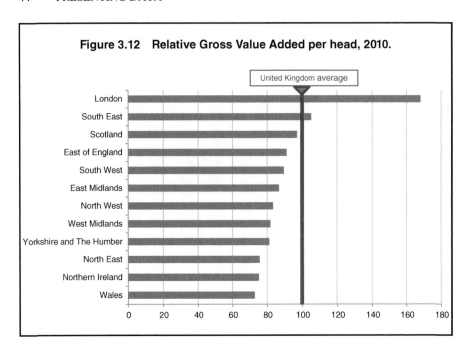

Figure 3.12   Relative Gross Value Added per head, 2010.

Principle 3.5: Arrange the chart to maximise the horizontal text.

One further issue with bar charts – that often results in the scale adjustment illustrated in Figure 3.6 – is redundancy. This is evident where much of the chart does not illustrate anything but commonality. Figure 3.13 shows the proportion of the population in different districts in Malawi that are male. This is the chart automatically produced by Excel, given the data.

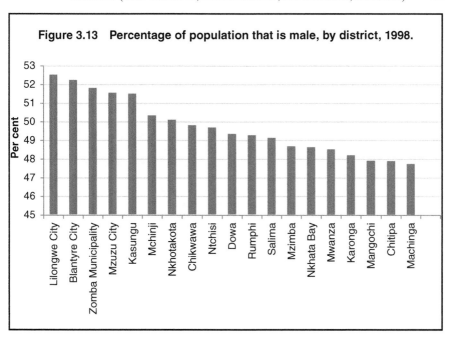

Figure 3.13   Percentage of population that is male, by district, 1998.

As previously noted, users try to understand the meaning of the chart without looking at scales or footnotes. The visual impression of Figure 3.13 is that the proportion of males in the cities (left-hand side of the chart) is over twice the proportion of that in Machinga, a rural area. Basically the men are moving to the cities to find work and get money to send home to their families in the rural areas.

Now let us force Excel into putting the scale in full and turning the chart through 90 degrees so we can read the names of the areas a little better.

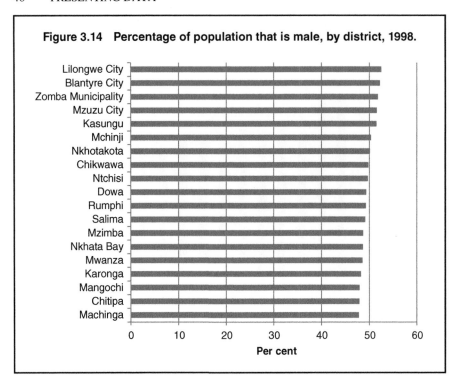

Figure 3.14   Percentage of population that is male, by district, 1998.

This chart, as Figure 3.13, does illustrate that the districts with the larger proportions of males in the population are the four major cities and that only seven of the areas have a proportion of more than 50 per cent.

This redrawn chart is properly drawn with the length of each bar of correct relative size to the others. The problem is that more than 90 per cent of the chart is redundant and it becomes difficult to see the differences between the adjacent bars. So, whilst it is correctly drawn from the data, it is not too illustrative of the data and the differences between districts. We need to remove the redundancy in some way and allow the user to see clearly the differences in the data.

Further, we can consider some 'added value'. With these data, the 'added value' could be the average proportion of the districts shown.

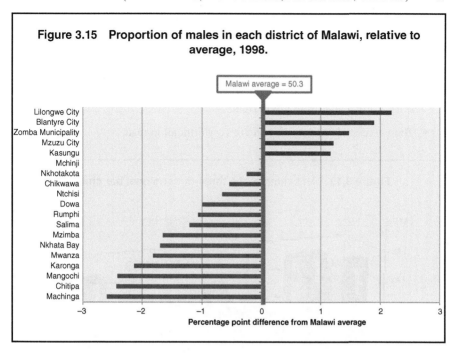

Figure 3.15 Proportion of males in each district of Malawi, relative to average, 1998.

It is easy from this chart:

- To read the district names;

- To identify that five areas – mainly cities – have above-average proportions of males;

- To note that the range is within +/−3 percentage points of the Malawian average.

This chart not only has no redundancy but also gives a greater definition of the differences compared with Figure 3.14. Similar issues arise with graphs – see Section 3.8.

One of the most difficult presentations for users to understand the scale is in three-dimensional bar charts. The following example, Figure 3.16, not only has the issue of the third dimension, but also of overlapping bars – making it even more difficult for the user to assess relative size. The poor quality of the chart is not from the reproduction – it was this poor in the original – so the user would have had difficulty reading the text and scale as well as the information in the chart itself.

The issues with this example are pretty obvious:

- The way the chart was produced has resulted in non-straight lines for the grid lines and axes;

- The producer thought the differentiation of the two bars to be difficult – so added up or down arrows to show movement between years; the most similar bars (third group from the left) do not have an arrow – and are not equal – so should have;

- The later data bars are to the left of the earlier data bar – and we like to tell the story – and read – from left to right;

- The y-axis scale and x-axis text are very difficult to read.

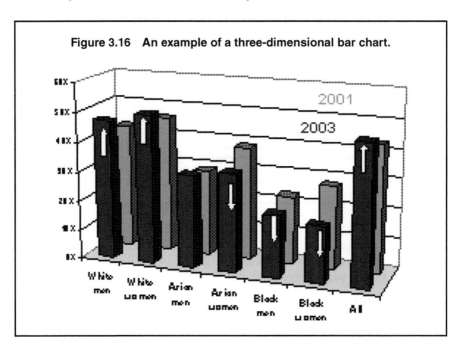

Figure 3.16   An example of a three-dimensional bar chart.

But the producer of this chart did know how to do better! In the same document, indeed on the same page, Figure 3.17 was shown. This is much more readable and certainly easier to interpret – but still not quite as good as it could be. Two changes could be made to improve it: first, remove the repeated '%' symbol from the y-axis scale and add a specification 'Per cent'; secondly, soften the gridlines into a pale grey. The need to add arrows to the bars has gone; the text is readable and the data in the right order.

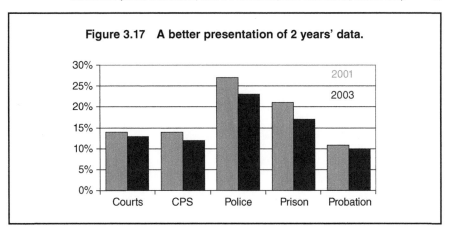

Figure 3.17   A better presentation of 2 years' data.

3.4   Difference between bar charts and histograms

Technically, two differences exist between bar charts and histograms: first in the type of data presented and second in the way they are drawn.

Bar charts display 'categorical data', that is, data that fit into categories – such as the districts of Malawi in Figure 3.15. Each category is independent of the others.

Histograms are used to present 'continuous data', that is, data that represent a measured quantity where, theoretically, the numbers can take on a value in a range, for example, age, height, weight or wages. The data collected would be grouped to display in a histogram.

Drawing the histogram is slightly different from a bar chart. The bars for each category in a bar chart are usually separated by spaces, as in Figure 3.15: where data for 2 years are adjacent in a bar chart, no space would be inserted between the bars for each year but space is inserted between the groups for each category's bars, as in Figure 3.17. For a histogram, the bars are placed next to each other to show the continuous nature of the distribution of the variable.

Two examples of histograms are given as Figures 3.18 and 3.19. Again, as we have already seen, it is sometimes better to have the chart turned through 90 degrees so that the labels can be read more easily. Whether it is or not also depends on the aim of the chart. If the aim is to get a sense of the magnitude of the bars, it may be better to have the bars in a vertical position – unless the labels become difficult to manage in which case, the chart with the horizontal bars would be easier to understand.

In Figure 3.18, the aim was to show the changing dynamics of the population in:

- The impact of increased births;
- The net inflow of university students and
- The fairly steady decline in the numbers in each 5-year age band from 60 to 64 years.

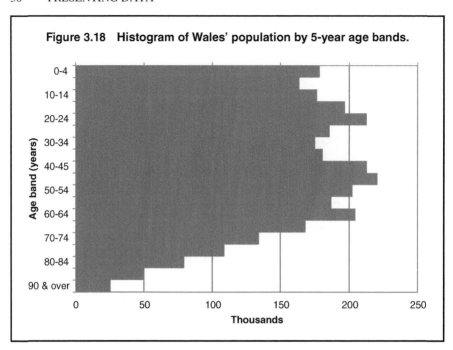

Figure 3.18   Histogram of Wales' population by 5-year age bands.

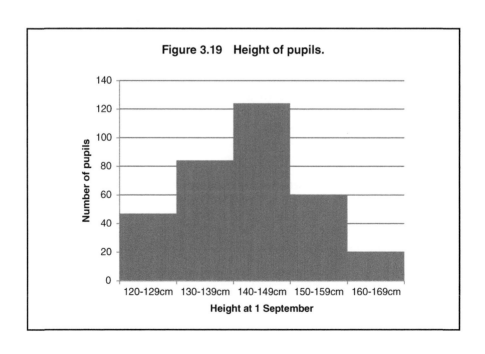

Figure 3.19   Height of pupils.

3.5   Pie chart principles

Many believe that the prettier a chart, the better it is. But is it? Probably not. The main consideration should be whether the presented chart communicates the information matching the aim. So many examples seen in publications today have multicoloured charts, sometimes in three dimensions and not representing the data effectively to communicate the message contained in them. Sometimes it is appropriate to have more than one colour – but the rationale for the many colours must be understood. Similarly, the basic principles of each type of chart must be applied – or the resultant chart is unlikely to reflect the data well, making the understanding of the message much more difficult.

For pie charts, the fundamental principle is that the divisions of a pie add to the whole, that is, 100 per cent. The technical term for the individual parts of the pie is segment. The examples used in this section are similar to published examples.

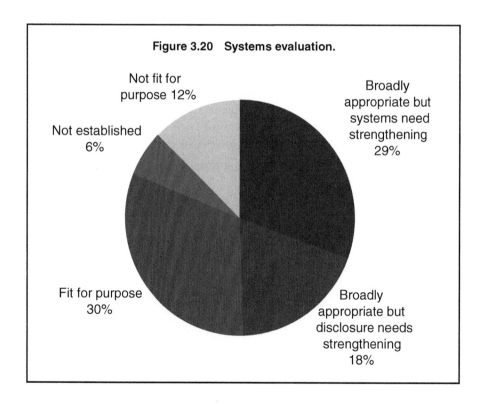

Figure 3.20   Systems evaluation.

For those quick in additions, you will spot that the sum of the percentages given in Figure 3.20 is not 100 but only 95. The title did not have a footnote marker but notes were given under the original. The first note indicated that 5 per cent of the systems had only recently started operating and thus it was 'too early to judge the strength of the controls in place'. But the chart has a segment 'Not established' so they could not be judged either! A letter was sent to the head of the organisation pointing out the basic principle of a pie chart and the sum of its constituent parts. The response was that they had noted why the 5 per cent were missing. It is actually not any harder to draw a chart where the constituent parts sum only to 95 per cent in Excel – but it would present the user with much more difficulty in understanding the data! Also look at the order of the information: to assist the user in the interpretation of the chart, the order would be better as:

Fit for purpose
Broadly appropriate but system needs strengthening
Broadly appropriate but disclosure needs strengthening
Not fit for purpose
Only recently started
Not established

The other issue here is the use of colour. Only one of the segments was in a different colour – grey – but that would have been more appropriate for the segment 'Not established'. This chart could be improved by having three colours: green for established systems that are reasonable, red for the systems that were not fit for purpose and grey for those that could not yet be assessed (those recently started and those not established systems). The result would be Figure 3.21.

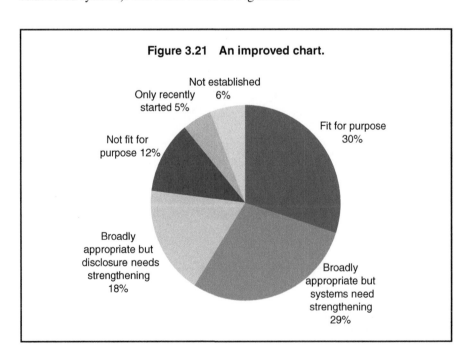

Figure 3.21   An improved chart.

The next pie chart (Figure 3.22) is an American example measuring recidivism from different offences, that is, the proportion of people convicted of certain offences that reoffend. It is quickly obvious that the basic principle of pie charts has not been observed: the total of the percentages in the segments is 329.1. The data here do not form the division of a whole (100 per cent) and, in fact, should either be shown as five different pie charts (each showing the proportion reoffending and the proportion not reoffending) or the chart needs to be completely rethought and probably drawn as a bar chart! If one chose the five pie charts option, it would be quite difficult to discriminate visually between the relative sizes of the reoffending proportions as they are quite similar.

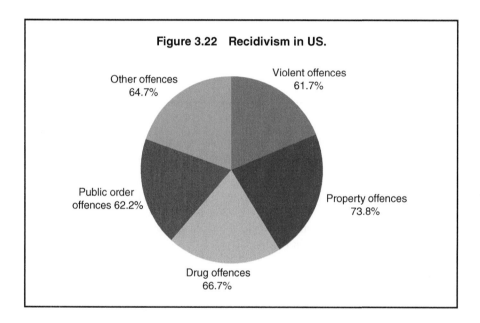

Figure 3.22   Recidivism in US.

Other offences 64.7%

Violent offences 61.7%

Public order offences 62.2%

Property offences 73.8%

Drug offences 66.7%

Principle 3.6: Ensure the sum of the percentages in a pie chart add to 100.

The aim of the chart is obviously to look at the different rates of recidivism in the prisoners in the United States. The only way such a chart could be used is where each segment of the pie is in proportion to the number of individuals who have reoffended with the rates shown to the side. This, however, would be quite confusing to any reader and would require a significant written description – negating the use of the chart.

So, if the basic aim is to show the different rates of recidivism across the offence groups, a bar chart would be the appropriate type of chart, as in Figure 3.23.

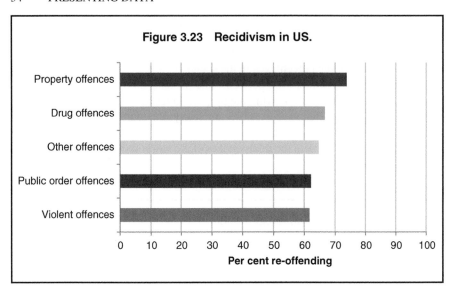

Figure 3.23   Recidivism in US.

Note that here we have the 0 to 60 common and thus not adding to the story. We could add the average for all types but that would not help much.

Again thinking around the possibilities, we could calculate the average rate of recidivism and show the data relative to this figure. To be technically correct, the average should be a weighted average of the percentages but those data were not readily available. So, for illustrative purposes only, the arithmetic average of the percentages has been used as a base figure in Figure 3.24.

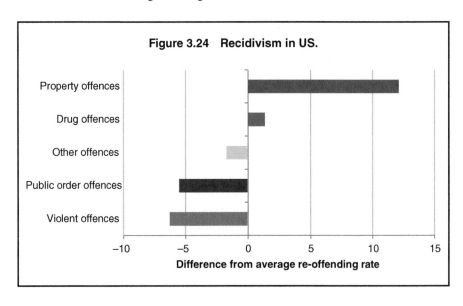

Figure 3.24   Recidivism in US.

Essentially, Figure 3.24 puts a magnifying glass on the area of Figure 3.23 that is the most interesting. The aim is clearly met and the users can understand that the people committing property offences are much more likely to reoffend than other offence groups.

> Principle 3.7: Ensure that a pie chart is the most appropriate type of chart to meet the aim.

3.6   Issues with pie charts

Six major issues with pie charts appear frequently in those presented by the press or in reports, particularly business reports:

- not starting the top segment at 12 o'clock and progressing clockwise;
- not thinking about sorting the data;
- using three-dimensional pie charts;
- having too many segments;
- the poor use of colour and
- asking users to compare different sizes of pie charts (this issue is covered in Section 3.10).

Let us consider each of these in turn.

Why should a segment of the pie chart start at the 12 o'clock position? It is back to the very basic principles we all learnt in school. We were taught to tell the time from the start of the hour and clockwise around the clock in quarter or half hours. So, for the user, it is easier to judge the proportion of the pie in the first segment if it starts from the 12 o'clock position.

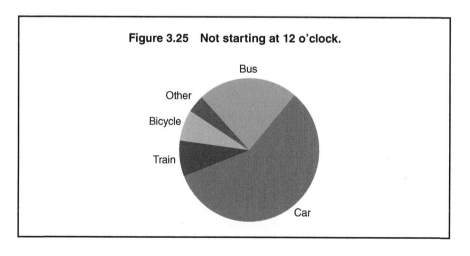

Figure 3.25   Not starting at 12 o'clock.

Consider the top segment in Figure 3.25: is that above or below a quarter of the pie? It is actually just less than one quarter – but hard to see from this chart. So what is the aim of the chart? If it is to monitor the reduction in car usage to get to work, it would be appropriate to have that segment starting at the 12 o'clock position and going clockwise. If the aim is from the bus transport authority and they want to monitor the uptake of bus transport to work, it would be appropriate to have the bus segment starting from the 12 o'clock position. Let's just see the latter in Figure 3.26.

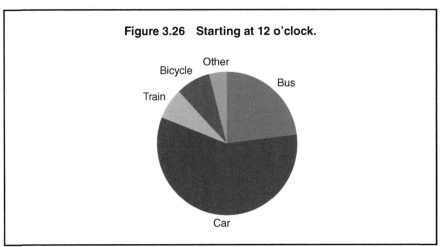

Figure 3.26   Starting at 12 o'clock.

It is now easy to see that the bus usage is just under a quarter – the message one wants to give. However, if the changes over the time periods of measurement are small, that is, of the order of one or two per cent, then another pie chart side by side with this one would not allow the user to distinguish adequately any change – and the producer would have to think of some other way of presenting the information (e.g. going back to a table of changes?).

In the literature, an alternative start for the segments, at the 3 o'clock position, is sometimes suggested. For me – and the majority of users – a start at the 12 o'clock position is easier to measure from and interpret. This is because all of us are used to looking at a clock and measuring time past or to the 12 o'clock position.

It is clear that the aim should drive the chart's design and bring clarity to the presentation of data. If the chart is produced without an aim, it will not generally be productive in communicating a message.

Principle 3.8: Always start the main segment of the pie chart at the 12 o'clock position, progressing clockwise.

The second issue is the sorting of the segments of the pie chart. If one is looking at the policy of travel to work and seeking to write a report on the current position and then describe policy to change the position, it would be appropriate to produce a chart with the aim of showing the current position in a pie chart. For this aim, the data would be sorted by size with the largest segment being the first from the position of 12 o'clock. Figure 3.27 illustrates this: note that the percentages of the various modes are included – these help the user to distinguish the actual data where segment sizes are similar (like for train and bicycle).

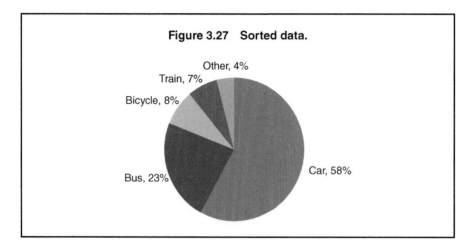

Figure 3.27   Sorted data.

Sometimes it is not appropriate to sort the data by size because there is a more important ordering. For example, if one is looking at a certain characteristic by age, then it is more important to keep the data and ordering of segments in age order. Let us look at a distribution of the married population of England and Wales by age in 2010 (Figure 3.28).

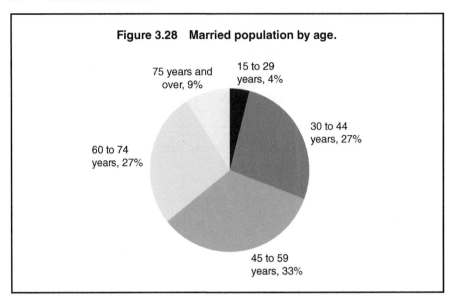

Figure 3.28   Married population by age.

It would be nonsense here to order the segments by size as the order would be: 45 to 59 years, 60 to 74 years, 30 to 44 years, 75 and over and 15 to 29 years.

Principle 3.9: Where appropriate, sort the segments of the pie into size order.

The third issue with pie charts is the inclusion of depth of pie by adding a third dimension. Many times the reason for presenting pie charts in three dimensions is that they are 'more realistic' with depth and they 'look better'. The difficulty here is that the third dimension actually makes it much more difficult to assess the relative sizes of the segments because of the angle of tilt and the blurring of the relative segment sizes by the additional depth in the front segments of the pie. Let me illustrate this with an example.

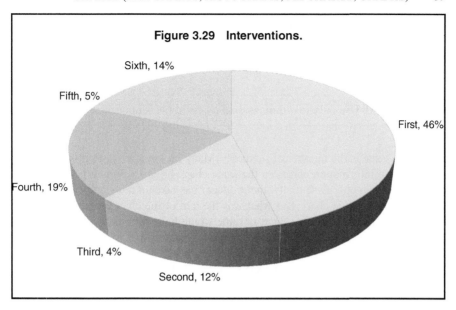

Figure 3.29   Interventions.

If we look at the pie chart in Figure 3.29 without looking at the data labels, the second intervention proportion of the pie looks considerably larger than the sixth intervention segment. In data terms, it is the other way round: similarly for the comparison between the third and fifth intervention segments.

In order to reinforce the argument, I have counted the pixels in each of the segments and set them, as percentages of the whole, in the same table as the actual data: these are shown in Table 3.2.

Table 3.2   Interventions.

	Actual percentage from data	Percentage of visible area
First	46	43
Second	12	22
Third	4	6
Fourth	19	17
Fifth	5	3
Sixth	14	9

So the eye leads the user to one conclusion but the wrong one! Comparing the actual percentages, it is clear that the segments to the front of the pie chart are given greater weight than those without any apparent depth.

Thus, when asking a user to judge the differences in data, do not ask them to adjust what is visible for tilt and depth when doing a comparison. In plain English, this would be expressed as keep two-dimensional charts in two dimensions so that the chart represents the data effectively and accurately.

Principle 3.10: Use only two-dimensional charts in two-dimensional media.

The fourth issue is the number of segments. Many presentations of pie charts try to get too much information across in the same chart. Figure 3.30 is a good example of how not to do one. Essentially, the producer has taken a table and converted the whole table into a pie chart. But what was the aim of the chart? If it really was to show each of the values, a bar chart or table would have been better. If the purpose was to show the relative sizes of Agriculture and Manufacturing segments, a pie chart with just four segments would have been better, combining everything from Financial activities to Other services into one segment, labelled 'Other groups'.

One obvious issue in this chart is the labelling of the segments: the descriptions are long and lines to their segments cross text. The maximum number of segments should be 6 for clarity. If the data has more categories than this, consider grouping and recheck the aim. Another possibility, as in this case, is to do a bar chart instead – or just a simple table with one column of percentages (with the percentage symbol in the column heading, not against each number)!

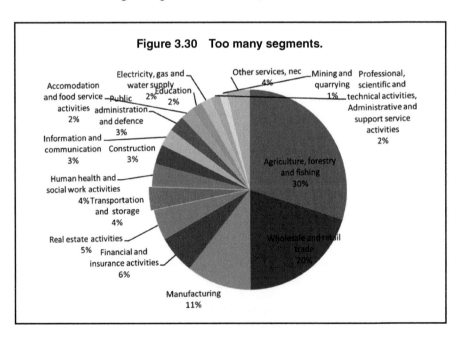

Figure 3.30   Too many segments.

Principle 3.11: The maximum number of segments in a pie should be 6.

The next issue is colour. Two aspects need to be considered. First, bright (primary) colours are dominant in the brain of the user. So putting primary colours in a chart can draw the user's brain to them as opposed to giving equal consideration to each part of the chart. Let us re-present Figure 3.31 with a couple of primary colours in the segments.

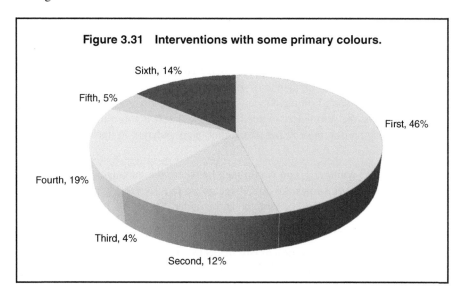

Figure 3.31   Interventions with some primary colours.

Sixth, 14%

Fifth, 5%

First, 46%

Fourth, 19%

Third, 4%

Second, 12%

The user's brain is drawn to the red and yellow segments and the ratio between the second and sixth segments appears to have changed, possibly because the brain is more focussed on the red and the top part of the second intervention segment, not its depth.

Principle 3.12: Avoid use of bold primary colours in charts.

The second aspect of colour use is where multicoloured charts are shown when, in fact, the categories are part of the same variable distribution. In the mode of travel to work chart (Figure 3.27), the categories are independent.

Whereas, in Figure 3.28, we are just looking at proportions of the same variable (age of the married population): and, in this case, the segments should be of different densities of the same colour. Figure 3.28 is one where the age classification has trumped the distribution and the same colour is used from deepest to lightest. In this case, as with all other charts, the producer must go back to the aim and check the chart against the aim and ensure that the age distribution is more important than the proportion of the age group that is married.

Principle 3.13: Use shades of the same colour when presenting proportions of one variable.

If the data could be put into a bar chart, use different colours for the segments; if the data should be put into a histogram, use shades of the same colour.

As an example of where the age classification is not the most important, let us think of a chart's aim to show which age group needs targeting with the messages of AIDS/HIV. In Figure 3.32, the chart has segments of the pie for each age group, their size relative to the numbers known to have HIV (the prevalence). The shading could then represent the incidence of HIV in the last year and the colour of the segments could be sorted on incidence – with the highest having the darkest shade. Checking against the aim here, we need to ensure that it is more important to have the age ordering predominate rather than either the population in the segments or the incidence.

This and the next younger group would then be where the message targeting would be directed. The figure shows the 20–29 age group to have the highest incidence in 2011 so the advertising could be targeted at the under 20 and 20–29 age groups.

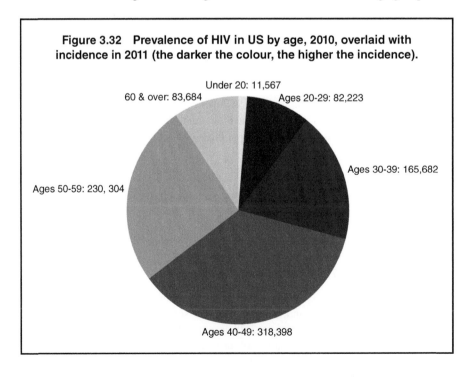

Figure 3.32   Prevalence of HIV in US by age, 2010, overlaid with incidence in 2011 (the darker the colour, the higher the incidence).

Under 20: 11,567

60 & over: 83,684

Ages 20-29: 82,223

Ages 30-39: 165,682

Ages 50-59: 230, 304

Ages 40-49: 318,398

3.7   Graph principles

A graph is a line showing the relationship between two or more variables. The majority of the graphs that are commonly seen show the relationship between two variables – and one of those is often time. The axes for the variables are at right angles. The user is looking for two things in a graph: the height of the line (sometimes called the level) – or series of lines – and any pattern shown (could be a seasonal pattern or a trend or a combination of both). For every graph where we are trying to show the size of the variable, the scale on the y-axis should start at zero. Where the data are indexed (to show changes from a start point), the value of 100 is equivalent to the zero on a standard graph.

For some data, the aim of the chart is to show movement over time with very restricted change over the period, for example, the value of a performance indicator or a rate such as the total period fertility rate. Hence, in these cases, the concentration is not on the height of the graph or its height relative to another graph on the same chart. Some examples are given towards the end of the next section.

When time is one of the axes, it is usual to show this as the x-axis horizontally across the page with time progressing from left to right. It may seem obvious to state that the intervals on the x-axis should be equal for the same time period. If this is not done, the apparent change in some periods will appear greater or less than those in other periods – but will not reflect the changes in the data. Illustrations of this issue are given in the next section.

A graph is meant to tell a story, usually over time, and thus – in the Western world – should be prepared to be read as we normally read, that is, from left to right across the page.

3.8   Issues with graphs

Let's consider two different graphs, Figures 3.33 and 3.34. Both show the population of towns over time. The visual impression of the graphs in these two figures is:

(a) that the population of Town A was a new town in 2003 and grew rapidly and

(b) that Town B was already established in 2003 and did not grow so quickly.

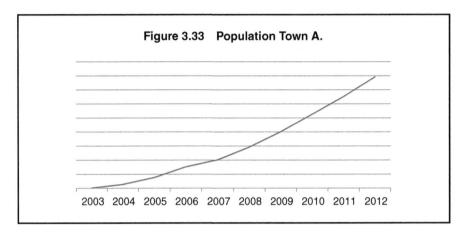

Figure 3.33   Population Town A.

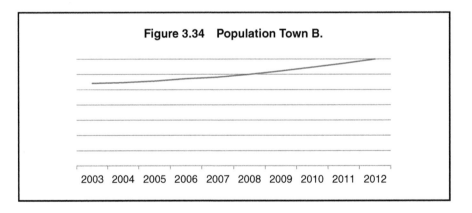

Figure 3.34   Population Town B.

Visual impressions are gained from just the central picture. Indeed, it is commonly understood that the user will look to the picture first and seek to understand what the picture is telling them and ignore scales and notes. Therefore, if the picture does not represent the data well, the user will be misled and could get the wrong message.

These last two graphs are now re-presented as Figures 3.35 and 3.36 with scales on the y-axis.

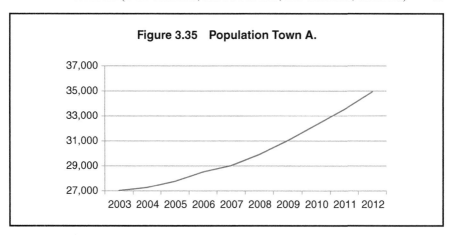

Figure 3.35   Population Town A.

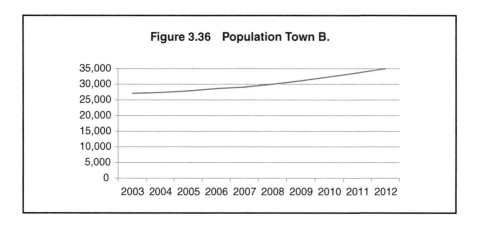

Figure 3.36   Population Town B.

It is now obvious that Town A was not a new town in 2003. Indeed the data are the same – but just shown on different *y*-axes. The correct presentation is in Figure 3.36 but the scale has a lot of redundancy – too many zeroes that are not needed. We can call the scale 'thousands' to lose those, alter the background colour and make the gridlines less obtrusive to produce Figure 3.37.

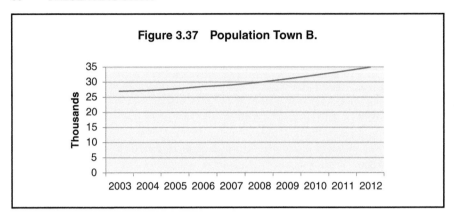

Figure 3.37   Population Town B.

Does this chart meet the aim? If the aim was to show the steady growth in population over time, it does.

Principle 3.14: When graphing simple numbers, always start the y-axis at zero.

Now let us do a comparison between two charts. Figures 3.38 and 3.39 represent immigration to and emigration from the United Kingdom.

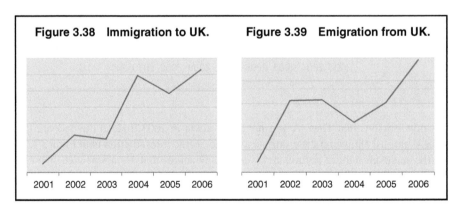

Figure 3.38   Immigration to UK.   Figure 3.39   Emigration from UK.

The message given to the user who looks only at the picture of the chart is that the trends are similar with a similar start, an upward trend in both and a slightly higher close in the emigration data. When originally presented, the scales were on the right of each chart and had both different start and end values – as in Figures 3.40 and 3.41.

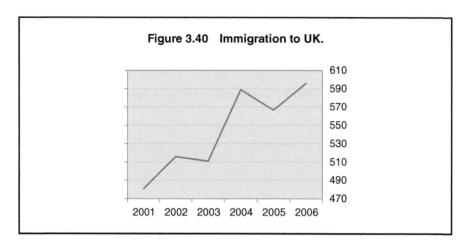

Figure 3.40   Immigration to UK.

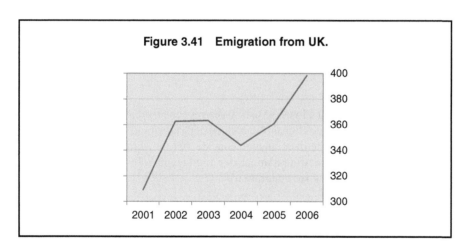

Figure 3.41   Emigration from UK.

It is obvious now that, even though both display an upward trend, immigration is far higher than emigration. The user here will have to do a lot of interpretation of the charts to get the real message. But the producer could easily have done something to assist.

To help the user understand these numbers properly, it is necessary either to draw them both to the same scale, side by side, or to put both parts on the same chart – as in Figure 3.42.

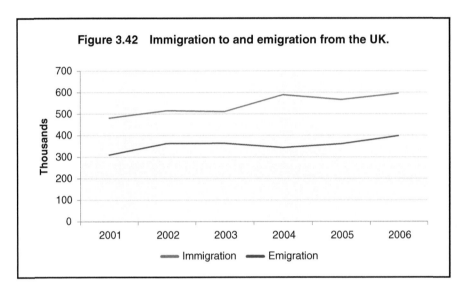

Figure 3.42   Immigration to and emigration from the UK.

The latter is preferable as it allows the user to see very clearly that the immigration was always higher than the emigration through the period shown and, because the gridlines are helpful in assessing the data, anyone can see that the number of immigrants is roughly one and a half times the number of emigrants in 2006 (6 gridlines high for immigration compared to 4 for emigration).

Principle 3.15: When asking the user to compare two data series on graphs, use the same scale for both series.

One other issue that occurs with graphs is when the producer wants to graph two series of very different levels, for example, the population of countries. Some have drawn graphs with different scales on the left and right of the graph for the different countries. Again, it is important to look back to the aim for the chart. Let us consider a chart to show changes in the populations of four countries over time. A simple graph of the numbers, using Excel defaults, produces Figure 3.43.

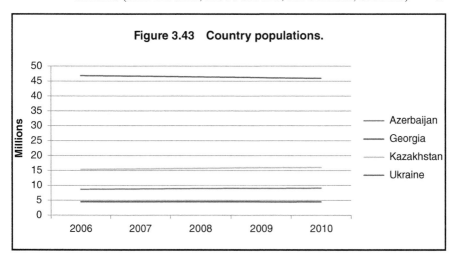

Figure 3.43   Country populations.

With this chart, it is clear that Ukraine is significantly higher in population terms across the whole period. Also it appears that Ukraine is losing a little population and Kazakhstan is growing a little. But the user has additional work to do because the legend is not in the same order as the lines appear on the graph: Excel does allow these to be changed quite easily – but it is the producer's responsibility to examine the chart before release to see if it meets the aim and if anything can be improved. Simply trying to understand the graph would indicate that the countries are in a different order to the legend. Let us change the legend order, add a background and change the gridlines to give Figure 3.44.

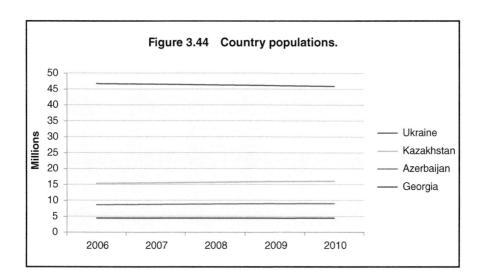

Figure 3.44   Country populations.

So how can we see what is really happening to these countries' populations? The simple technique we can use is called indexing. This puts all of the data in the start year equal to 100. Then each of the subsequent data items is calculated relative to that baseline. As an example, the first two population figures for Ukraine are: 46,667,646 and 46,509,400.

We set the base figure (the first one) for Ukraine equal to 100.0. The second indexed figure is then:

(second figure divided by first figure) multiplied by 100.0 which is (46,509,400 divided by 46,667,646) multiplied by 100.0 = 99.7

And the third indexed figure is calculated as:

(third figure divided by first figure) multiplied by 100.0 etc.

Now, the 'base' figure for the graph is 100, not zero in a conventional graph. Figure 3.45 displays this. Note that the y-axis scale can go below 100 implying that the population of Ukraine was falling over the whole period. From this chart, it is clear that the populations of Azerbaijan and Kazakhstan increased each year and that of Georgia fell in the first 2 years, steadied and then rose in 2010 to a figure higher than that in 2006. This does meet the aim – to clearly see changes in population: however, it does not show the relative sizes of the populations. If the latter is important, the best solution would be to include a simple summary table of the populations – appropriately rounded.

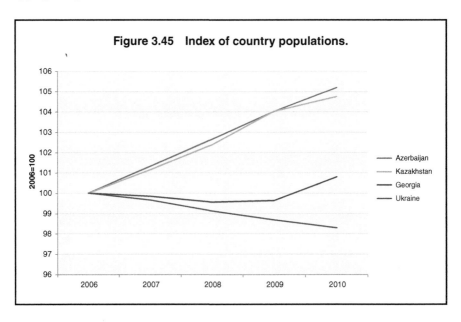

Figure 3.45   Index of country populations.

> Principle 3.16: When graphing series of different magnitudes, consider converting the data to index numbers before graphing.

At times, the data being graphed are in a very narrow range. The aim of the chart would then not be the presentation of the absolute values of the data but the changes in the data over time. Let us consider two examples: the monitoring of a rate or the regular measure of activity against a target.

For the first, let us consider the total period fertility rate (the number of children born to a cohort of women aged 15–44). Over the recent period, this rate has remained within a band of 1.5 to 2.0. So charting this with a zero base for the y-axis produces Figure 3.46.

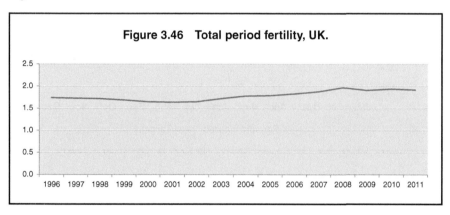
Figure 3.46   Total period fertility, UK.

If the aim was to show the level over time, this chart would adequately provide the information. But the aim was to consider the changes over time and in this case, 80 per cent of the chart is then redundant as no data appear in four of the five sections. Further, it is not easy to note some of the movements in the data or the data values over time. This could then be redrawn with a different scale to give Figure 3.47.

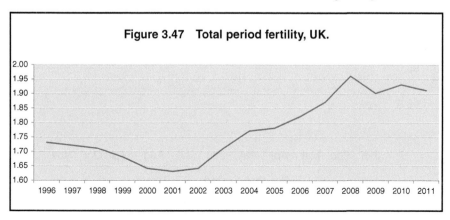
Figure 3.47   Total period fertility, UK.

This resulting chart indicates that the rate has changed significantly within the relatively narrow range and it is clear that three distinct trends can be clearly observed.

Another similar example is when we consider target data for a specific operation. A target value is set and one is concerned not with the level of the chart but with the data relative to the target. Using a zero base for the y-axis produces Figure 3.48.

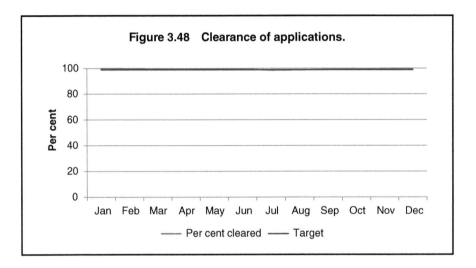

Figure 3.48   Clearance of applications.

The red and blue lines are practically indistinguishable. Changing the y-axis scale to show just the relevant part, 98 to 100, we have Figure 3.49.

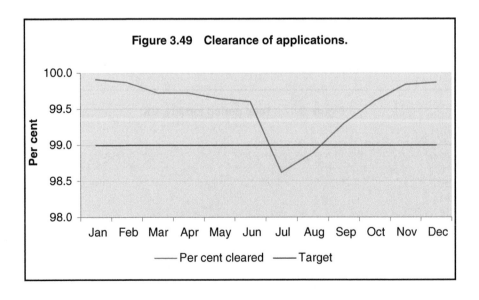

Figure 3.49   Clearance of applications.

This clearly shows that the per cent cleared dropped below target in two of the months but not drastically. Considering these data and Figure 3.49, we could make the message clearer by using a bar chart as in Figure 3.50.

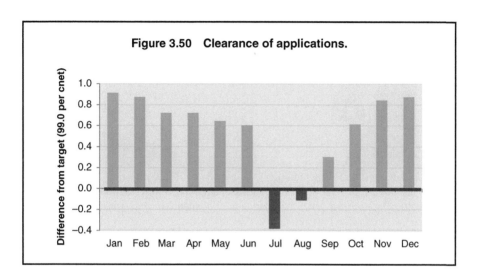

Figure 3.50 Clearance of applications.

With Figure 3.50, the red colour is very striking and would instantly show that a problem had occurred as soon as the datum for July was available. So, from this example, it is not just necessary to ask if the proposed chart meets the aim but also whether the one proposed can be bettered. The question the producer should ask is 'Can the data be displayed in another way which presents the message in a clearer and simpler way?'

Charts for new reports should have a significant part of the development time devoted to them. Where charts are for an update of reports, the producer should not only check back to see that the chart still meets the aim but also ask whether the message could be made clearer through a simpler or better type of chart or presentation.

> Principle 3.17: Always ask if the message is clear and whether the chart could be simpler or better.

Another issue with graphs is the abuse of the standard understanding of the progress of time along the x-axis. Such abuse means that the chart does not represent the data and their underlying message. Two examples are provided to show this. A figure similar to Figure 3.51 was first presented in a German magazine, *Wirtshaftswoche*, in 1988 seeking to show the demise of the British car industry. The dramatic fall at the start of the chart shows, at first glance, that the British car industry was almost wiped out in one period – with the car shown about to crash through the x-axis.

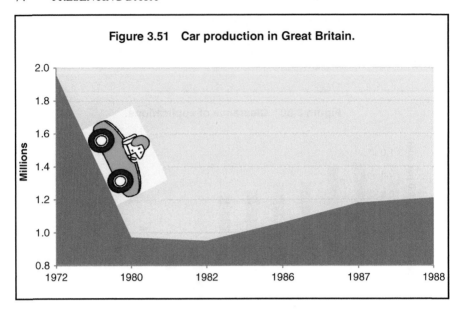

Figure 3.51   Car production in Great Britain.

However, two errors are apparent in the construction of the chart: first the y-axis scale does not start at zero and so much of the production is below the x-axis; secondly, the years on the x-axis do not have the same time interval between them. The first interval is eight years, then two, then four followed by two single years. In the original, the decimal separator used was not consistent: on the y-axis it was a comma for all apart from the '0.8' when it was a full stop!

The compression of the first 8 years into one interval on the chart makes the drop significantly steeper that it was in reality and thus indicating the wrong message. In technical terms, the gradient (dy/dx) of the first section of the line shown in Figure 3.51 is –0.99 instead of –0.12. If the chart were drawn with these two errors corrected, we would have Figure 3.52 – which has a very different message.

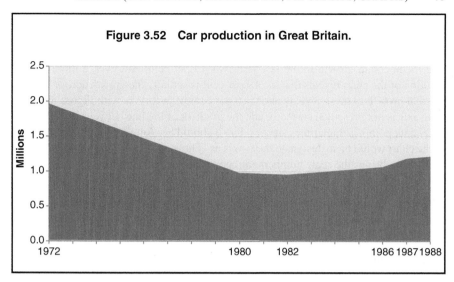

Figure 3.52   Car production in Great Britain.

Another poor example is concerned with the message given to the user. Here, in a redrawn chart, Figure 3.53, the producer showed the growth in the number of people registered with credit unions on top of the chart itself. The y-axis scale was started at 20,000 – acceptable as the aim was to show the growth over the years.

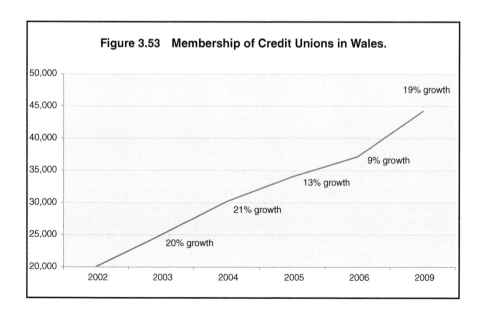

Figure 3.53   Membership of Credit Unions in Wales.

A quick glance at the chart shows greatest growth in the first couple of years, then a slowing down followed by a sharp increase in the last period – to which the user's eyes are automatically drawn. This last growth rate, left in the mind of the user, is that membership growth in the last year was 19 per cent. However, a more careful scrutiny of the chart reveals that the 19 per cent is actually the growth across 3 years: the real growth in these 3 years is only 6 per cent per year! As with Figure 3.51, the time axis is not presented correctly and the gradient of the line between the 2006 and 2009 data points is three times steeper than it should be. Another small improvement to the chart would be to designate the y-axis as 'Thousands' and hence lose 21 zeros – which also brings the scale numbers into normal understanding range.

Principle 3.18: Ensure the data points are appropriately spaced so as not to misrepresent the message.

Finally, for this chart, one has to ask if a simple table would have been as effective as the chart. Table 3.3 simply gives the membership numbers and the growth year on year.

Table 3.3   Credit unions in Wales.

	Membership (Number)	Year-on year growth (%)
2002	20,000	. . .
2003	24,900	20
2004	30,100	21
2005	34,000	13
2006	37,100	9
2007	39,300	6
2008	41,700	6
2009	44,200	6

With the numbers and growth rates now in columns, it is easy for most to be able to note that membership has more than doubled in the 7 years of data shown but that the growth rate peaked in 2004. The aim for the chart was not written down – so one can only surmise that the producer was trying to show growth in membership: but was it the overall growth since 2002 or the recent growth? If the former, Table 3.3 can be simplified even further to Table 3.4.

Table 3.4   Credit unions in Wales.

	Membership (Number)	Growth (%)
2002	20,000	...
2009	44,200	121

A final caution when using automatic generation of graphs: always check the scale and ensure that it is easy to understand and appropriate. Excel has two decimal places as default. One example by a government organisation had the y-axis scale from £0 to £2 million in half-million-pound intervals with the numbers written as:

£2,000,000.00
£1,500,000.00
£1,000,000.00
£500,000.00
£0.00

Chapter 1 urges the producer to think about the user's appreciation of the numbers and, to communicate, this would have been better as:

£ million

2.0
1.5
1.0
0.5
0.0

and with the descriptor of £ million by the side.

Another poor example seen is given in Figure 3.54 which is a made-up chart (used with permission) of Internet access by continent with the y-axis scale going from 0 to 1.1 billion in 0.1 billion intervals. The difficulty for the user was that the numbers on the scale were not presented as here with one decimal place of billions, the final number being shown as 1100000000 – without thousands separators.

Figure 3.54   Change in the number of Internet users by broad geographical area.

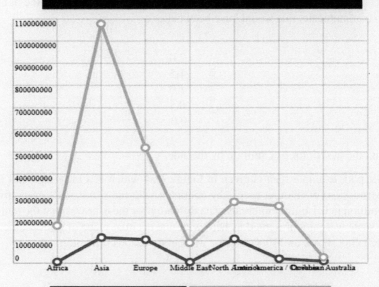

INTERNET PENETRATION 2012 - GLOBAL TRENDS

There were an estimated 2,405,518,376 internet users worldwide for mid-year 2012 (June 30, 2012), This represents about 34.3% of the population worldwide and a 566.4% growth compared to 2000

www.internetworldstats.com

Internet Users Dec. 31, 2000   Internet Users, June 30, 2012

Other issues with this chart include:

- The numbers in the quote box at the top are unrounded;
- The use of a percentage as high as 566 is not understood by many: it would be better to say 'From December 2000 to June 2012, the number of Internet users had grown by more than five times';
- The text of the x-axis overlaps;
- The y-axis scale is left-justified, without thousands separators but needs to have the numbers more concise through rounding;
- The type of chart is completely wrong! It is shown as two graphs but should be a bar chart with two bars for each continent.

So, redrawing this in Excel and sorting the continents according to the number of users in 2012, we have Figure 3.55.

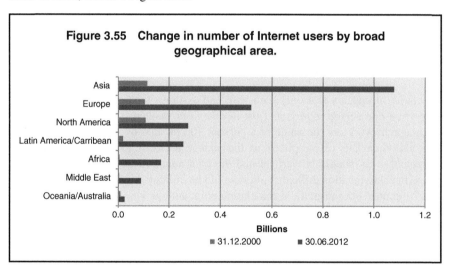

Figure 3.55   Change in number of Internet users by broad geographical area.

Principle 3.19: Ensure the basic principles of data presentation are followed for the scales on charts.

Principle 3.20: Make sure the correct type of chart is presented.

3.9   Pictogram principles

This chapter started by considering a simple pictogram. It was easy to understand and related data to a picture which sought to capture the imagination of the user. As with all charts, the important starting point is a clear aim and the important closing point is checking that the resultant chart meets the aim. Clarity, simplicity and understandable are keywords here. If a pictogram takes a long time to interpret and understand its message, it is the wrong one!

Figure 3.56 does not meet any of the principles of the key words – and is incorrectly drawn!

Figure 3.56   Voting in elections.

Voted in national and local election,
national only, local only, **neither**

On clarity, it appears that only men vote and assumes that all can work out what proportion one person represents (just in case your mathematical ability requires a calculator to work out the value of a person, 100 divided by $18 = 5.56$ per cent!). This also then fails on simplicity as the change between voting classifications is horizontal – but should be vertical and the continuous run makes the comparison across the classification difficult. And it would be difficult for any user to understand it in 2 minutes – because of the calculation point and the way it is set out (ask if the proportion voting in national only is the same as those who voted in neither).

Nothing is wrong with using a pictogram for this data – it is just that it could be done better. Let us look at the improved Figure 3.57.

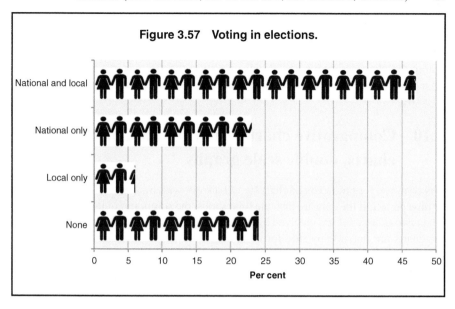

Figure 3.57   Voting in elections.

It is now clear that the proportion voting in national only is less than the proportion who voted in neither. The pictogram has both men and women shown and the user can reasonably easily see the percentage of the population in each category – without having to do the calculation with lumps or parts of 5.56 per cent!

Pictograms can have appropriate symbols in the chart. So, at the beginning of the chapter, we used a money bag; in Figure 3.57, we used people and Figure 3.58 uses cars – since the subject is the availability of cars in households.

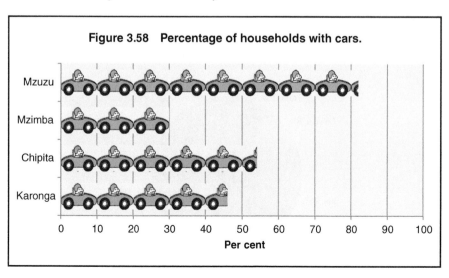

Figure 3.58   Percentage of households with cars.

Many newspapers now use coloured charts and pictograms to illustrate their stories. Some are well done but many take considerable thought and time to understand and interpret the meaning. If the user cannot understand the message quickly, they will ignore the chart and move on – totally negating the effort in producing the chart. So a clear, simple and understandable chart will both get the message across and be effective.

3.10   Comparative charts: Multiple pies, multiple bar charts, double scale graphs

It is common for producers of charts to tell a story with more than one series of data or time period. If the data are just one time series, the solution is quite easy. However, when two time series are involved in the same chart, issues can arise. Similarly, when pie charts are put side by side for comparison, we expect the user to be an expert in comparing up to six segments: some comparative pie charts seen are of different sizes – adding a further complication. This section examines some of the issues.

The first principle of pie charts is that the total (100 per cent) is split into segments representing the elements of the whole. In drawing a pie chart, it is normal not to consider its radius since we do not assess the overall quantity by the total area of the pie. However, if the producer wishes to show both the split of the total across 2 years and some growth in the total, it would be necessary to understand the formula for the area of a circle. This is dependent on the square of the radius (area $= \pi r^2$, where π is a constant and r is the radius of the circle). So a circle with a radius of one unit has an area equal to 1π square unit and a circle with a radius of two units has an area of 4π square units.

Taking Figure 3.27, let us renumber it as Figure 3.59 and take it as the base for 1.2 million travellers into a city each morning.

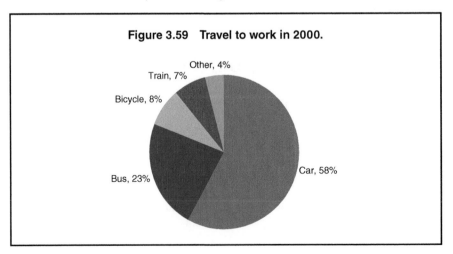

Figure 3.59   Travel to work in 2000.

Other, 4%
Train, 7%
Bicycle, 8%
Car, 58%
Bus, 23%

Let us suppose that the number of travellers into the city increases to 1.8 million in 2010 and thus we need a larger chart to represent the increase. The ratio of the travellers is:

$$1.2:1.8 \text{ which is } 1:1.5$$

If we used this ratio for the radii, the relative sizes of the circles would be

$$\pi(1 \times 1) \text{ square units}: \pi(1.5 \times 1.5) \text{ square units}$$
$$= 1\pi \text{ square units}: 2.25 \ \pi \text{ square units}.$$

Using this ratio for the radii is thus wrong. We need the square root of the ratios of the traveller numbers, that is,

$$\sqrt{1}:\sqrt{1.5} \text{ which is } 1:1.2247$$

So the ratio of the radii in the comparable charts has to be in the ratio 1 to 1.2247 as in Figure 3.60.

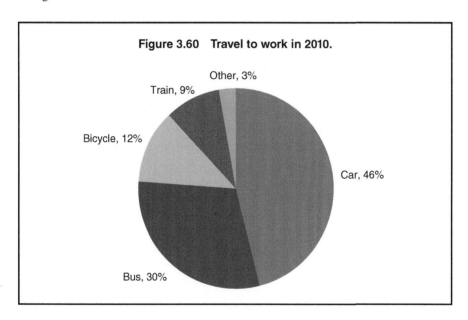

Figure 3.60   Travel to work in 2010.

How many professionals, let alone normal users, can see the difference in size of the pies and conclude that the travellers have increased by 50 per cent? And then we ask them to judge the change in the sizes of the segments. All too complicated for the ordinary user. With these data, it would be better to have a statement about the increase in travellers and to have a comparative bar chart to show the differences in the modal split – as in Figure 3.61. So the aim of the chart would be to show the differences in the proportions of travellers using each mode of travel to work. This would be used alongside the policy objective for the area which is to persuade travellers to leave their cars at home and go by public transport, bicycle or on foot.

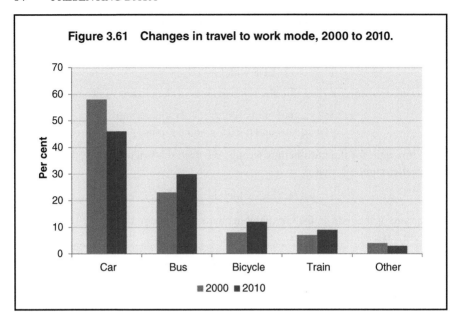

Figure 3.61   Changes in travel to work mode, 2000 to 2010.

From this chart the user can easily see the changes in the modes of travel between the years. In addition to this, a simple sentence could be:

> The number of people travelling to work each day has increased from 1.2 million in 2000 to 1.8 million in 2010.

Figure 3.61 is a multiple bar chart – of the simplest variety since there are only two series being displayed. Some can be seen with 10 series together in the groups of bars. Figures 3.62 and 3.63, representations of Figures 3.3 and 3.4, have four bars in a group and even here one has to ask what the aim of the chart is. Earlier in the chapter we discussed how having one word different in the aim changes the resultant chart significantly.

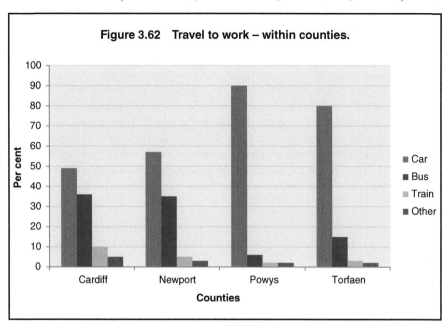

Figure 3.62   Travel to work – within counties.

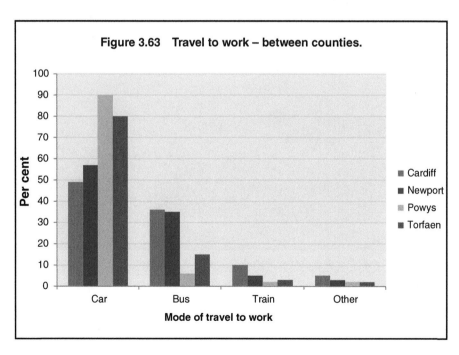

Figure 3.63   Travel to work – between counties.

Essentially, having drawn the chart, the producer has to identify clearly what they are asking the user to compare. Next, check whether this is simple or the user needs to compare individual bars which have many other bars in between.

In the previous paragraphs, the bars discussed have been side by side – so at least the user has a common baseline from which to measure differences. Where the user has more difficulty is when the data are presented as stacked bar charts. Because the baseline for all parts other than that at the bottom is not level, it can be difficult to assess what is happening to the individual parts. Such stacked bar charts, as Figure 3.64,[2] can look colourful but are hard to interpret apart from the trend in the bottom section and the total. If the aim of the chart is to identify changes in any element other than that at the bottom of the chart, it would be better to consider if even a table would be better.

Principle 3.21: Ensure the information in the chart can be quickly assimilated and understood.

Figure 3.64   Number of people entering central London in the morning peak, 1978 to 2011.

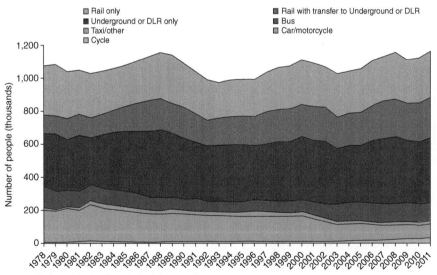

Source: TfL Group Planning, Strategic Analysis.

Note that if this was split into seven individual bar charts, it would be essential to have all using the same scale for the y-axis so the visual impression of the size of each mode would be comparable. The aim of these charts would be to see the comparable change in each type of transport. If the report later considered just one of the forms of transport, it may then be appropriate to change the scale on the y-axis to reflect the data being presented for that mode.

Finally in this section, it is important to address double-scale graphs, where more than one graph line appears on a chart, with independent *y*-axis scales. Such charts have their place for professional users who can readily assimilate the subject matter and have a good understanding of the use of graphs and their reading. Obviously, the producer is trying to show that a relationship exists over time between the two graph lines. As an example consider Figure 3.65 looking at the number of people killed on roads in Wales and the number of vehicles licensed in the principality to be on the road.

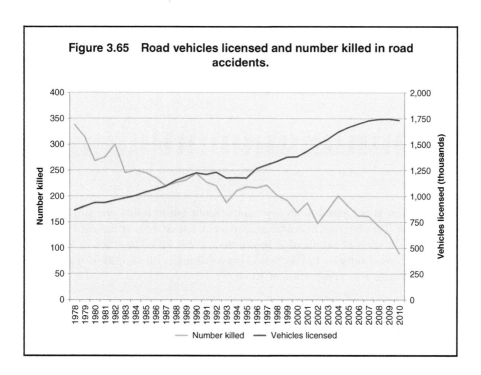

Figure 3.65   Road vehicles licensed and number killed in road accidents.

It is clear that the trend in vehicles licensed is upward and that the trend in the number of people killed on roads in Wales is downward. However, the relative change is almost impossible to deduce as it is very dependent on the scales chosen for the *y*-axes. The user also has difficulty understanding two scales: in this chart another complication is the different units of the scales as one is in thousands whilst the other is in individual units. Surely the aim of such a chart would be to show that roads are getting safer. Is this the best way to meet the aim? Probably not. Why not just have a single graph of a ratio between deaths and the number of vehicles licensed as in Figure 3.66?

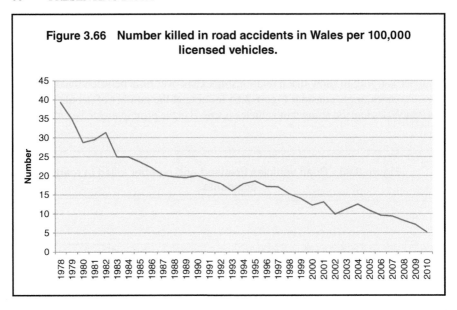

Figure 3.66   Number killed in road accidents in Wales per 100,000 licensed vehicles.

Here the user is not asked to do any computations but to follow the single line of information. The change is also dramatic: so much so that the user can easily identify that in the 32 years of information shown the number killed on roads in Wales per 100,000 licensed vehicles has decreased from just under 40 to 5. Most will also be able to calculate 5 as a proportion of 40.

Principle 3.22: When tempted to use double-scale graphs, ask if they can be simplified to assist with the understanding of the message.

3.11   Graphics

Today's newspapers are full of graphics trying to explain or illustrate the data being presented. Sadly, many do not represent the data clearly for the ordinary user and many exaggerate differences which mislead the reader. And not only do issues with the graphics occur but also with numbers presented. Two examples will suffice here – but astute readers of the newspapers will often be able to spot others on a regular basis.

For the Olympics in 2012, *The Times* on 4 August 2012 carried a graphic depicting two events that a British athlete was to do that day together with data on the events. The events were the long jump and the javelin. The data in the first half of the graphic, for the long jump, were:

Personal bests:

Jessica Ennis	6.51 m
Tatyana Chernova	6.63 m
Natallia Dobrynska	6.82 m
Olympic record	7.27 m
World record	64.64 m

From these data, it could be said that either the Olympians are not good at the long jump or that the person holding the World record cheated and used a jet pack! The answer is actually simple: the figure given for the World record is wrong.

In the second part of the graphic, the data for the javelin were:

Personal bests:

Jessica Ennis	47.11 mm
Tatyana Chernova	49.25 m
Natallia Dobrynska	54.49 m
Olympic record	55.7 m
World record	64.64 m

Now it becomes clear: the ridiculous figure for the World record of the long jump is actually the World record for the javelin. But poor Jessica Ennis appears to be no good at this event as she has a personal best of much less than half a metre! Looking at the data, could it be that the Olympic record is actually 55.70 m, the zero being dropped because of the program it is held in? This is one of Excel's issues: look at the data and force two decimal places for all of the data to be consistent and get proper alignment of the data.

The second example comes again from *The Times*. In April 2011, a special report for Holy Week reported the number of Christians in 2010 and the expected number in 2050 for a selection of countries. In the following table are some of the figures presented for 2050:

USA	301,962,000
China	225,075,000
Brazil	222,469,000
D R Congo	179,432,000
Nigeria	139,000,000
India	113,800,00
Ethiopia	112,046,000

The data were presented in a proportional space typeface which goes against the principle of lining up the digits in columns of the same value. See Principles 1.4 and

1.5 in Chapter 1. Following these principles, the hundreds of millions digit for D R Congo would be aligned with the tens of millions digit of Nigeria. The absence of a zero in the units' column for India shows that the data were not checked effectively before publication. And, finally, why do we have so many digits in the numbers? All of the table could have been given rounded to millions – which would have made the table much less difficult for those not too at ease with numbers: this would have removed six digits from all of the numbers leaving just 3!

The moral of these presentations is that data in graphics MUST be checked by the responsible data owner before publication. That will then achieve correctness, consistency and clarity. In quality terms, it is said to be completing the quality circle – essential to achieve the highest quality of output. For any producer who passes his or her output either to a media team who then prepare a press release or to a website team for Internet publishing, it is essential that the final output – as seen by users – is checked by the producer, preferably before release to users.

Within earlier parts of this chapter, we have dealt with the proper presentation of data in charts. Many examples can be found where the graph or chart has not been drawn correctly to represent the data well. Written aims and adequate checking will improve the charts presented.

Principle 3.23: With graphics, complete the quality circle and do a final check of the data in the graphic against the source before publication and check the final graphic against its aim.

3.12   Three-dimensional charts

Within this chapter, examples have been given of three-dimensional charts: a bar chart (Figure 3.16) and a pie chart (Figure 3.29). Other examples abound in reports and the press. Many are included because 'they look pretty' or 'they appear technically advanced'. But why are charts included in documents? Surely they are to appeal to the visual appreciation of the brain to communicate data in a spatial as opposed to a numerical way. If a researcher presented data that had been skewed as with the elements of three-dimensional charts included above, users – and their employers – would question their integrity. But to do the same with charts appears not generally to be frowned on, presumably because the users or employers are not always aware of what is happening or the misrepresentation that occurs.

The basis of the above comments is that the producer, as part of the aim for a chart, wishes to represent the data in a graphical way which mirrors the relationships in the data themselves.

The major difficulties for the user of such three-dimensional charts are:

- It is difficult to adjust all charts back to two dimensions in the mind to assimilate properly the relative sizes of the bars on a bar chart or the segments of the pie;

- For bar charts, the user cannot easily read off the gridlines to read the value;

- The third dimension distorts the relative sizes of chart elements – and thus the elements do not reflect the actual differences in the data.

So, where the data are two-dimensional, charts also should be two-dimensional. If the data are three-dimensional, for example, in surface mapping within topology studies, then the appropriate chart can be three-dimensional. Note, however, that many users will have difficulty interpreting such charts so it is essential that the aim of the chart is clear and that the resultant chart is compared well with the aim to see that the aim is met.

> Principle 3.24: Only use two-dimensional charts for two-dimensional data.

Summary of principles: Charts

General

Principle 3.1:   A chart must represent the data clearly and accurately.
Principle 3.2:   Before starting to create a chart, have a clear and specific written aim.
Principle 3.5:   Arrange the chart to maximise the horizontal text.
Principle 3.12: Avoid use of bold primary colours in charts.
Principle 3.20: Make sure the correct type of chart is presented.

For bar charts and histograms

Principle 3.3: Ensure that the visible areas of the bars are in proportion to the data.
Principle 3.4: Arrange the elements of the chart to maximise understanding of the information.

For pie charts

Principle 3.6:   Ensure the sum of the percentages in a pie chart add up to 100.
Principle 3.7:   Ensure that a pie chart is the most appropriate type of chart to meet the aim.
Principle 3.8:   Always start the main segment of the pie chart at the 12 o'clock position, progressing clockwise.
Principle 3.9:   Where appropriate, sort the segments of the pie into size order.
Principle 3.10: Use only two-dimensional charts on two-dimensional media.
Principle 3.11: The maximum number of segments in a pie should be 6.

Principle 3.13: Use shades of the same colour when presenting proportions of one variable.

For graphs

Principle 3.14: When graphing simple numbers, always start the y-axis at zero.

Principle 3.15: When asking the user to compare two data series on graphs, use the same scale for both series.

Principle 3.16: When graphing series of different magnitudes, consider converting the data to index numbers before graphing.

Principle 3.17: Always ask if the message is clear and whether the chart could be simpler or better.

Principle 3.18: Ensure the data points are appropriately spaced so as not to misrepresent the message.

Principle 3.19: Ensure the basic principles of data presentation are followed for the scales on charts.

Principle 3.21: Ensure the information in the chart can be quickly assimilated and understood.

Principle 3.22: When tempted to use double-scale graphs, ask if they can be simplified to assist with the understanding of the message.

Graphics and three-dimensional charts

Principle 3.23: With graphics, complete the quality circle and do a final check of the data in the graphic against the source before publication and check the final graphic against its aim.

Principle 3.24: Only use two-dimensional charts for two-dimensional data.

Notes

1. Two similar phrases attributed to Frederick R. Barnard in *Printer's Ink*: 'One look is worth a thousand words', in *Printer's Ink*, December 1921, and 'One picture is worth ten thousand words' in *Printer's Ink*, March 1927. Also Arthur Brisbane in speaking to the Syracuse Advertising Men's Club, in March 1911, used the phrase 'Use a picture. It's worth a thousand words'.

2. Used by kind permission of Transport for London. The original was Figure 1 in the 2013 edition of *Central London Peak Count (CAPC), Travel in London Supplementary Report*.

4

Numbers in text

For many users, the presentation of the numerical message in either tables or charts is difficult to understand with the loss of the fundamental message. This may be a product of poor presentation which can be corrected by following the guidance of Chapters 2 and 3: the result should be tables and charts that are both well-presented and from which it is easy to extract the message. For other users, it may be that they have an aversion to tables and charts and prefer the written word with numbers included in the text.

The purpose of including numerical information in any document is to describe a situation or to present a message. Without the numerical information, the description or presentation becomes vague. So we have statements like

Immigration has increased massively.

Many people die from obesity.

The work had a major impact on residents.

Banks will flood into the city.

But what does 'massively', 'many', 'major' or 'flood' actually mean? Are they backed up by the numbers? How can the user put the statements into some kind of scale to know whether they should be concerned about the information being presented or not?

When numbers are described effectively, the message in the data is clearly understood by the user. But what does the producer have to do to the numbers to make them understandable? We begin by looking at the basic principles of including numbers in text applying the principles introduced in the first chapter to illustrate changes from poorly understood information to memorable and understandable information. Later we consider some of the other issues relating to the inclusion of numbers in text.

Presenting Data: How to Communicate Your Message Effectively, First Edition. Ed Swires-Hennessy.
© 2014 John Wiley & Sons, Ltd. Published 2014 by John Wiley & Sons, Ltd.

4.1   Numbers written as text

It is most commonly suggested that numbers up to 9 or 10 should be put in words in text and not numbers. This is sensible – but many of the numbers we wish to use in messages or reports are much greater than 10 and we need to know how to present them. In the introduction, the producer was urged to consider the presentation of information against a checklist of four words.

These principles should be applied to the numbers we wish to present in text.

4.1.1   Correct numbers

Nothing frustrates users more than finding that the information presented is incorrect. Further, the reputation of the presenter or, worse, their organisation is damaged. Trust for, and reputation of, the individual and organisation are hard to acquire but easy to lose. Who wants a reputation of being not trusted to present correct data? This can simply arise from missing that last stage of checking information presented in a graphic from information sent to the designer (see Section 3.11). Or it could be that a typing or rounding error has occurred as in the following which appeared in a recently received report:

> Cumulative revenue over the first 5 years estimated to be between $5760k and $839k.

Fortunately with the statement, several tables were provided and I could deduce that the first number was incorrect as it appeared to be. Here the first amount should be $576k.

Principle 4.1: Check the numbers in the final presentation to ensure they are correct.

4.1.2   Clear numbers

In Chapter 1, we noted that most people do not handle numbers easily. Hence the more that can be done to make the numbers clear and understandable the better. How do we understand 11.016 million? Is that clear? The digits without the decimal separator actually relate to thousands so the number could have been written as 11,016 thousands. But how many of us would read the number and remember all of the digits? We would prefer to think of the number as just over 11 million. Within this heading it is also necessary to think how a user would understand the data presented. How many users really understand what a trillion is? Yet the national debts of countries are measured in these large units and the numbers issued as if everyone would understand: but few do!

So, in order to communicate to the user, we have to think of a way to bring the numbers into the conscious range of the user's understanding. In the United Kingdom, the national debt can be translated into the equivalent of £40,000 per household. In the United States, the Federal debt at the time of writing was about $16,738,158,460,000 which was $51,297 per person.[1] Obviously, these numbers for the United States would be better as $17 trillion and $51,000. With per-household or per-person numbers, everyone can appreciate the numbers relative to wages or house prices.

Recalling the example from Chapter 1 giving summary information about Wales, we could give the population in 2012, which totalled 3,074,100, and the number of sheep in the country, recorded as 8,898,200. But how many of us after reading these numbers in a document would remember the numbers a few pages later? But if we say:

Example 4.1 ....

Population of Wales is around 3 million
and
Number of sheep in Wales is around 9 million.
That is, three sheep for every person in Wales!

These numbers will be remembered for a long time.

> Principle 4.2: Ensure the message from the numbers is clear.

....

4.1.3   Concise numbers

Within Chapter 1, we looked at effective rounding of data. This ensures that the numbers presented are as concise as possible when we are trying to compare numbers. The US Federal debt when compiling this chapter was noted as $16,738,158,460,000:[2] this is not presented concisely. On its own, we should just say around $17 trillion. However if we compare this figure with that for the 2010 financial year, using effective rounding, we should give the figures as

	$ trillion
2010 Financial year	11.9
September 2013	16.7

Nevertheless, most would translate these numbers into $12 trillion and $17 trillion for the ease of understanding.

Many producers put numbers in press releases with decimal digits when, in reality, no decimal digits are needed to understand the information. Let us look at two simple examples:

Example 4.2 ....

The US budget deficit came to $680.3 billion[3] ... down from $1.09 trillion.

If the first figure was the only one quoted, we should give $680 billion – the '.3' is irrelevant. But this quote has breached one of the other rules – consistency of units when asking users to compare numbers. So this would be better written as:

The US budget deficit came to $0.7 trillion ... down from $1.1 trillion.

....

Example 4.3 ....

The proportion ... jumped from 19.9 per cent in 2003 to 27.3 per cent this year.[4]

This was meant to be a key message to the public. Do they want such precision? Probably not and they can understand the effectively rounded information as ...

The proportion ... jumped from 20 per cent in 2003 to 27 per cent this year.

It is easy to understand that the producer has spent a lot of time collecting, processing, analysing the data before presenting them but do the numbers presented communicate quickly and easily?

Principle 4.3: Make the numbers in text as concise as possible.

....

4.1.4   Consistent numbers

To help the user understand a message, the presentation of numbers should be consistent. The presentation of numbers within a table was discussed in Chapter 2. Here, within text, the numbers should be consistent: where comparison is required, with the same number of decimal places and the same units. Many tables of data are presented with differing numbers of decimal places: this can then cause the use of such numbers in text as some writers do not want to change the source numbers. It is also inadvisable to present numbers for comparison in units of millions and thousands: the conversion of numbers to different units before comparison is too complex for the ordinary user. An example of this, given as a bullet point in a press release is shown in Example 4.4.

Example 4.4 ....

11.048 million tests were completed in 2007/08, an increase of 563 thousand (5.4 per cent) on the previous year. Between 2002/03 and 2007/08, there was an increase of 1.386 million (14.3 per cent).

So, here, not only are the first two numbers in different units, the increases quoted are also in different units! To gain a quicker understanding of the information, we should put all of the data into millions and apply some rounding as:

> 11.0 million tests were completed in 2007/08, an increase of 0.6 million (5 per cent) on the previous year. Between 2002/03 and 2007/08, there was an increase of 1.4 million (14 per cent).

Even this is not as clear as it could be since we are asking the user with one bullet point to assimilate two different periods of change and to think of increases after the final number was given. So splitting the point into two and changing a little gives:

> The number of tests increased by 0.6 million (5 per cent) in 2007/08 to 11.0 million.
> Between 2002/03 and 2007/08, the increase in the number of tests was 1.4 million (14 per cent).

Consistency does not only apply to the numbers and units we present in text. It also applies to the descriptors of the data that we use in describing our text. One of the worst examples of this used three different presentations of financial year in the same paragraph:

2007/08   2007/8   and   2007–08

Similarly, if the term 'passenger journeys' is used at the start of a press release on travel, the writer should not then abbreviate this to 'passengers' later in the text. For example, most passenger journeys into a city in the morning are followed by a return passenger journey in the evening. London Heathrow airport's statistics for 2012 records 'Number of passengers arriving and departing in 2012: 70 million'.[5] But in the same section of the data, a line says that 26.0 million or these were transfer passengers who arrived and departed, usually within the same day. Note the different rounding in the same section of data presentation! So does this mean that 44 million different people went through the airport? No, of course not. The majority would be counted at least twice for when they arrive and when they depart or vice versa. But, in addition, some people will make multiple journeys and thus be counted many times. So the words around the numbers have to be clear in defining what the numbers relate to.

> Principle 4.4: Be consistent with decimal places, the units of numbers and descriptors when comparing numbers in text.

. . . .

4.2   Ordering of data

Within Chapter 2, the principle of how we read data over time in tables was established – time running left to right or down the table. This is based on the memory and retention of data. The last number we read is assumed by our brain to be the latest despite the words around the numbers saying something different. The statement below, quickly read, leaves the last and, by implication the latest, number as $169 million.

Example 4.5 ....

Profits in 2012–13 increased to $207 million, up from $169 million.

Do we expect the user to calculate the increase in profit? Is that significant?

We have to ask what the most important aim is in this communication. Is it the final amount of profit, the fact that profit increased, the amount of the increase in profit or the rate of increase? The two alternative proposals below have different aims but the same underlying numbers.

Profits increased from $169 million in 2011–12 to $207 million in 2012–13.

Profits increased in 2012–13 by $38 million (22 per cent) to $207 million.

The second statement gives more and clearer information, though it is possible to improve even this by separating the message from the numbers. This would give:

Profits were significantly up in 2012–13. They increased by $38 million (22 per cent) to $207 million.

Principle 4.5: Put numbers in sentences in time order.

....

4.3   Technical terms

The world around us abounds in technical language and acronyms. The technicians who write about their work often write so that they, or someone in the same field, would understand the information. But who is the ultimate user of the information and would they be sufficiently aware of the technicalities?

Consider a large government department: a researcher looking into a specific treatment for a problem set of IB houses writes a note for the head of the department about the steel structural sub-frame and the appropriate solution using a mix of named chemicals.

Here, the first issue is the use of an acronym: IB. This is not an abbreviation, as Google would suggest, for the International Baccalaureate. It is a technical term used for 'Industrial build' houses which are built in parts in factories and then the parts are erected on site. Second, most of us would not know that one type of these houses is constructed around a steel sub-frame. Third, neither any of us nor the head of the department would know about the chemical to be used.

Had the note been intended for the builder who was a specialist in dealing with problems in this type of house, it can be assumed that they would know about the technical issues and the chemicals to be used. So we have to consider who the user of the information will be. The Internet has made this task more difficult as we can never be sure who will be reading what has been deposited and what their level of understanding of the subject will be. This will be picked up in the next chapter.

Looking at a particular website, a press release started with:

> 2013/14 changes to QIPP comparators
>
> Following feedback from NHS and commercial partners, a new QIPP prescribing comparator has been introduced, and changes made to four of the existing comparators, effective from Q1 2013/14 data.

Here, we have two acronyms: NHS can be guessed by UK residents as 'National Health Service' and QIPP may mean something to regular users but not to the general public (who has access to the press releases).

The easiest way to help all users with regard to acronyms is to spell out the words on first use of the acronym with the acronym in brackets following. If a document is long and contains many acronyms, it is sensible, in addition, to include a page of acronyms within the introductory pages. In the Internet age, if one does not want to spell out the meaning even for the first use of an acronym, one can always hyperlink words to a pop-up explanation. Indeed, on the site from which the above example was taken, many hyperlinks appeared – but not to the acronyms!

Let us consider another, simpler, example where the presenter of the information used the questions on a data collection form as the basis for a statement in a press release.

Example 4.6 ....

> There were 26 fatal casualties from fires in Wales in the year ending 31 December 2007, up from 22 in the previous period.
>
> So what is a 'fatal casualty' to the ordinary user? A death.
>
> And how could we write 'in the year ending 31 December 2007' clearer? In 2007!
>
> We shall return to this example later.

....

Technical language is thus acceptable if only those who work in the area are to see the resulting report. However, if the report is for wider circulation, the language needs to be written at a level that the ordinary person can read it. Within an office context, a report for general circulation should be read by someone from a totally different area before release to ensure that the meaning of the report is clear. Perhaps the writer should ask, having completed a report for general public consumption, 'Could my grandmother understand it?' If the answer is 'No', then examine the areas that are not understandable and seek to change the sentence structure and descriptions so that the information and messages become easy to assimilate.

> Principle 4.6: Avoid technical language where possible but, if necessary, explain any technical terms used.

4.4   Plain language

Within some countries, movements have been founded to assess the readability of reports and releases from organisations. In the UK, the Plain English Campaign[6] is a model organisation that not only appraises written pieces but also runs courses on how to improve communication. In the United States, this has gone further with the establishment of a United States Access Board[7] to assess and improve communication from the federal agencies. This Board was established following the passing of The Plain Writing Act of 2010 which requires federal agencies to write 'clear Government communication that the public can understand and use.'

The main thrust of plain writing is to use short sentences in communication in a logical order, use active verbs and use lists if that would be easier for the user to understand. So, in all that is written, we should have the final user in mind and continue to ask the question as to whether all could understand what is written.

It is easy to get messages wrong and the use of plain language assists in ensuring that the message is clear and understandable. We must also be comprehensive. The message of the data must be clear and correct. Consider the next two examples where the wording is not quite correct: the first is a bullet point from a press release and the second a headline on a national statistical institute's website.

Example 4.7 ....

- The mean age at birth has increased from 26.2 years to 28.9 years.

  Here it would seem that the babies are waiting a long time to be born – and would be very well developed when born! Obviously a few words are missing to explain that it is the age of the mother not the person being born. So this would be better written as:

- The mean age of mother when giving birth has increased from 26.2 to 28.9 years.

  ....

Example 4.8 ....

- Business investment: 0.0% change in third quarter

  In this example, it would appear that the person writing the headline had taken the previous one that read '0.4 per cent change in second quarter' and just replaced the '4' with '0' without thinking then about how the final sentence read. So this one should read:

- Business investment: No change in third quarter

....

Principle 4.7: Use plain language when describing numbers.

Each of the principles described so far in this chapter should not be thought of in isolation since it is common for more than one of the principles to apply to the statements we write. For excellent writing about the numbers we have, we should combine the principles and seek the clearest possible message which will then be understood by the majority of users.

Let us consider the following examples taken from press releases.

Example 4.9 ....

■ The number of prescription items dispensed by community pharmacies in England and Wales in 2007–08 was 785.4 million. There was an increase of 40.4 million (5.4%) from 2006–07 when the figure was 745.0 million.[8]

Here we can see two issues: first, the rounding of the numbers and, second, the ordering of the numbers since the last one relates to the earlier period. Thus the message of the piece is not clear and the numbers not concise. Addressing these two issues, we could then have:

- Dispensing of prescription items by Community pharmacies in England and Wales is increasing.

- They dispensed 785 million prescription items in 2007–08, 40 million (5%) more than in the previous year.

....

Consider the following bullet point from a press release:

Example 4.10 ....

■ 196,476 new houses were built in 2013 compared with 239,172 ten years earlier.

With this example, the numbers need some rounding and the time periods are the wrong way round. Again, therefore, the message of the piece is not clear and the numbers not concise. Taking these points and redrafting, we could have:

- House building is declining.

- In 2003 some 240 thousand new houses were built whilst in 2013 around 200 thousand were built.

....

Finally, let us return to the example used at the end of the Technical language section. The full text of the bullet point is:

Example 4.11 ....

■ There were 26 fatal casualties from fires in Wales in the year ending 31 December 2007, up from 22 in the previous period.

With this example, we have already commented on the technical language used but also we have the number of deaths from the later period shown first. With these data, one also has to ask the question whether such a small difference between the numbers is significant: it could be the result of just one house fire in which four occupants died. So the suggestion here is:

- 26 deaths resulted from fires in Wales in 2007.

....

Principle 4.8: Apply number principles (from Chapter 1) to numbers used in text.

4.5   Emotive language

Describing numbers is an art. We are trying to communicate with people who may not have the greatest grasp on mathematics and who cannot easily compute differences and calculate percentages. Some reporting of data includes emotive terms that are not really supported by the data being presented but are designed to gain attention or stir emotions. Let us look at a few examples.

EU Public Debt rockets[9]

This example implies that the whole of EU public debt is rocketing. The article notes that EU Public Debt increased from 90.6 per cent of GDP to 92.2 per cent in a quarter (but the user has to pull the two numbers from paragraphs 1 and 3). The greatest change was in Ireland where debt increased by 6.6 per cent of GDP to 125.1 per cent. Now, a rocket goes to the sky, hopefully, in a vertical direction. As a minimum when using this term, one would expect the direction of increase in a graph to be fairly straight up – which this is not.

Factories In Britain 'Booming Again'[10]

Just reading the headline would suggest that the output of the factories is at its peak. Following down the article, it is clear that, whilst the growth in output and orders is increasing at a higher rate than has been seen for nearly 20 years, output at this time is below the pre-recession peak. The final quote in the article is that the manufacturing sector is on the road to recovery – a much better summary than the above headline.

Doctors risk sinking as low as the bankers (in terms of trust)[11]

The headline does not indicate how many positions in the professional trust table the doctors would have to go down to equal the bankers. Sinking implies a swift downward move in the table of trust but the article is only referring to the communication between the doctors and their patients.

Principle 4.9: Avoid emotive language in headings.

4.6   Key messages

None of us create numbers, tables or charts just for the fun of it. We are trying to get a message across to users (the audience) from a research study, a collection of data or acquired information. As with both tables and charts, we must start with an aim. That will set the various scenes and the producer then has to put the scenes in some order.

Historically, the conclusions of such research and study always appeared at the end of the work. In today's world, the Internet has changed the way we want to look at such reports. Now we are used to putting information in what is termed 'Inverted style' of writing. This is the format used in newspapers: headline, summary in first paragraph followed by detail.

For research studies, the ordering is by logical process order. However, for most such studies, it is now essential to have a summary or overview at the start. Users who only want to identify the 'so what' from a study will not wish to read a massive tome to get the message. Similarly, for those collecting data through surveys, the user does not want to wade through chapters on sampling, question and questionnaire design, fieldwork etc. before getting the overall message. These topics are important for the researcher but usually add little to the message and should be relegated to annexes.

From a communication of the message standpoint, we need to identify the key messages from the research or study. Then these need to be prioritised: by importance as defined by the aim. As an example, suppose the researcher is looking at different ways of processing a product with the aim of whole-process cost reduction. During the research, five different parts of the process were identified as improvable in cost terms. The key messages would be these five summarised and ordered with the highest cost-saving one first, then the next highest cost-saving message etc.

Principle 4.10: Limit the number of key messages to four or five.

Authors can be tempted to put too much text in key messages. But greater emphasis can be achieved through minimal wording. The one quoted earlier in this chapter is:

House building is declining.

That is certainly clear and concise and the user does not have to navigate figures or do their own calculations to identify what is happening with house building.

Another example used in my sessions on storytelling with official statistics comes from an annual update of casualties from road accidents. The table of 25 years of data showed me clearly that the lowest number of deaths in road accidents had occurred in the last year shown. Looking at the press release, this fact was not mentioned! Later, similar data have been used in Figure 3.63 – which still show the continuing reduction in road deaths. A little research identified that the number of deaths was actually not just the lowest in the 25 years of data shown but the lowest since records began. The first key message from these statistics should have read:

2010 saw the lowest number of people killed in road accidents in a year since records began.

For press releases on research studies and survey results, the key messages should be clear, concise and correct. These can be put into bullet points at the start: again, ordered by importance. Developing such clear and concise messages takes more time than just creating sentences with lots of numbers in them. If the key messages include any numbers, the principles noted earlier in this chapter on the presentation of numbers should be followed.

The last example is a clear case of where the producer should prepare charts as well as looking at the data tables before concluding what message comes from the data. The chart of just the number killed (as in Figure 3.65) would have shown a clear downward trend with the last point being the lowest.

Principle 4.11: Summarise key messages in clear, concise and correct sentences.

How many key messages should there be? We are talking of key messages so the number should be limited to four or five. One organisation limits the number of bullet points to four: a press release found on their website had four actual bullets but the sentences beyond the bullets numbered 10 which were not prioritised! On writing to the author to comment on the real number of points, the response was that only four bullet points were allowed by the organisation.

Principle 4.12: Order key messages in terms of importance.

The text that follows the bullet points should then go into more detail for each of the messages in order that they appear in the list of key messages – because they will have been sorted by importance.

Any major technical parts of a release should be placed in an annex: for example, the sampling strategy, the questionnaire design, the survey operation. With the Internet available now, such common features of surveys and the details of data collections should be available more easily. The news release could simply note that these are available and give a direct reference to them.

Summary of principles: Numbers in text

Principle 4.1:	Check the numbers in the final presentation to ensure they are correct.
Principle 4.2:	Ensure the message from the numbers is clear.
Principle 4.3:	Make the numbers in text as concise as possible.
Principle 4.4:	Be consistent with decimal places, the units of numbers and descriptors when comparing numbers in text.
Principle 4.5:	Put numbers in sentences in time order.
Principle 4.6:	Avoid technical language where possible but, if necessary, explain any technical terms used.
Principle 4.7:	Use plain language when describing numbers.
Principle 4.8:	Apply number principles (from Chapter 1) to numbers used in text.
Principle 4.9:	Avoid emotive language in headings.
Principle 4.10:	Limit the number of key messages to four or five.
Principle 4.11:	Summarise key messages in clear, concise and correct sentences.
Principle 4.12:	Order key messages in terms of importance.

Notes

1. http://www.usgovernmentspending.com/federal_debt_chart.html

2. http://www.usgovernmentspending.com/federal_debt_chart.html

3. Information from *The Times*, 31 October 2013, p. 45.

4. Information from *The Times*, 31 October 2013, p. 46.

5. http://www.heathrowairport.com/about-us/company-news-and-information/company-information/facts-and-figures

6. http://www.plainenglish.co.uk/

7. http://www.access-board.gov/

8. Extract from press release – General Pharmaceutical Services: England and Wales 1998–99 to 2007–08

9. http://www.theguardian.com/business/2013/jul/22/debt-crisis-eu

10. http://news.sky.com/story/1136080/factories-in-britain-booming-again

11. http://www.thetimes.co.uk/tto/health/news/article3858969.ece

5

Data presentation on the Internet

Even though the Internet has been around for publishing statistics since the late 1990s, many still do not understand how to use the medium for the dissemination of their data and messages. This chapter explains how the Internet dissemination of statistics developed and how new developments allow a different presentation of, and communication with, data. These new presentations both allow the producer to engage users effectively and also allow the user to determine clearly what is presented. A combination of these two changes has produced a step change in the possibilities available to the user so that they may understand what is happening in data over time. Stories buried in data come to life. Further, instead of the user having to take what the producer had provided, they can now investigate relationships in data and tailor outputs to their own specification.

Many advantages derive from the Internet publishing of data but also a few disadvantages. The main advantage is ease of obtaining data on many subjects from many places from almost anywhere in the world! For users who want data from several countries on the same subject, the issues of organisation and extraction come to the fore: different dissemination programs mean the user has to learn how each works to extract the information required. In the United Kingdom, for some subjects, a user has to learn four different programs for the extraction of data for the same subject; for other subjects, United Kingdom figures are published together.

However, if one can find the correct website for the country/subject data, the information can usually be extracted reasonably easily. The availability and accessibility of search programs help here. Those responsible for the dissemination via the Internet should check often the responses to a search for their organisation. For example, if

Presenting Data: How to Communicate Your Message Effectively, First Edition. Ed Swires-Hennessy.
© 2014 John Wiley & Sons, Ltd. Published 2014 by John Wiley & Sons, Ltd.

the producer was a national statistical institute, then the search should be for 'Official statistics "Country name" '. We expect to find the official site in the top three of the returned list – preferably at the top. Some of the countries' official statistics sites that I have searched for – even recently – have had their official site as far down the list as item 40 of the returned suggestions from the search engine!

Publishing data to the Internet opens them to more outside scrutiny and can quickly result in questions being raised helpfully to assist in the validation of the data. When data were only available in paper publications – and these were sold through many outlets – it was never possible to let all of the purchasers know of any errors identified until the next issue of the publication, though copies remaining in the publishing house at the time errors were found had a corrigendum inserted. With the Internet publishing, however, incorrect figures can be uploaded instantly and made available to all. So, if an error is identified, it can be corrected quickly and easily – though many major producers have procedures to follow in such cases requiring the issue of a specific notice about the error. In preparing for a course on presenting data, I found an error in data displayed in a dynamic chart on the website of the organisation to which the course was being delivered. A recreated version of the information shown is in Figure 5.1.

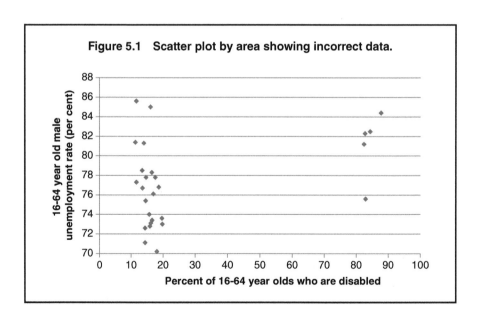

Figure 5.1   Scatter plot by area showing incorrect data.

This implies that the percentage of the population aged 16 to 64 years in certain areas which is disabled is over 80 per cent in five areas. Obviously, this is not true. A quick check confirmed this – and the erroneous data were the percentages of the population of the areas that are NOT disabled. So this is a simple data error.

Two courses were being given in the organisation on consecutive days and, having noted the error on day 1, the error was corrected by day 2.

Another major advantage is that instantaneous access to electronic data at publication time is available for all wherever they are in the world. In days gone by (and not too long ago), an update of one organisation's data had to be downloaded to magnetic tape and the tape couriered across the city to a consumer organisation.

One continuing debate in many statistical organisations is concerned with charging for data. The national statistical institutes of some countries have determined that, as the statistics are compiled from the public's and businesses' data, the statistics are a public good and should be made available free of charge. Others have budgets restricted on the premise that they will charge for at least some statistics and use the revenue to contribute to the office's running costs. Some allow access to the raw data free of charge but charge for added value services such as reports, analysis or tailor-made analyses or extractions.

The power of the Internet continues to increase year by year. For those websites that use text across all of the site and not graphics for links, the appearance of translation capability for web pages is a great asset and allows access to statistics of other countries with understanding. The translation of technical terms is not always precise in the resultant language but usually sufficient for the expert in a subject to obtain the correct data.

Few would argue that putting data onto the Internet is a bad thing. But it is not all plain sailing for the user. If the user wishes to collect certain statistics from different countries, each usually has its own way of organising and disseminating the data. For example, in the decentralised statistical service of the United Kingdom, each ministry decides the software tool for its dissemination: also, the four countries of the United Kingdom each have their own, different, software for accessing statistics. Take a subject like housing statistics in the United Kingdom and bring in an international context, say Netherlands, Luxembourg and Spain: four software systems for the UK data and then three other systems for the three other countries – so the problem increases. The user is very dependent on the site navigation, search engine and language choice of the sites to find the particular series of interest.

Since the Internet version of data always purports to be the latest, it is difficult to assess changes to data over time. Some organisations, for example, Way Back Machine,[1] some national libraries[2] and the United Kingdom's National Archive[3] are archiving versions of the sites but from these, it is difficult to assess changes in data especially since not all of the statistical pages or databases are archived. A few statistical organisations[4] have created electronic archives of data sets: some using specific versions of software have to be updated into newer versions so that the data can still be accessed. Few of these archives have complete copies of sites or the databases that are linked to them at specific periods: some are employing the techniques of the paper-publishing systems by adding cell notes or markers to data to indicate revisions.

5.1   The early years

When statistics were only being published on paper, the medium defined the possibilities and extent of data that could be presented. Many tables were presented badly and only a summary of the available data was ever presented. Very occasionally, attempts were made to bring together long time series of data. Some of these volumes were for a specific purpose – like the 100 years of British Labour Statistics or the two-volume set of Historical Statistics of the United States (bicentennial edition): from Colonial times to 1970:[5] these were welcomed but, in order to use the data, the user had to input the long series of data into a computer! The Welsh Office published a two-volume set of 700 pages of historical data for Wales with some data available back to 1570: these were soon out of print and the decision taken not to reprint them. However, as a means of preserving the access to the data, the tables were scanned, put into Excel and checked. During the checking process some minor errors in the data were noted and corrected. The result is that these data are now available both through the Welsh Government's[6] and the United Kingdom's Data Service's[7] website – and, obviously, open to all in the world who wish to look at the history of Wales.

The next stage in dissemination was to publish data on a computer disk and include the disk with the published paper volume. This was a great asset to the users of such data but the data quickly became out of date and such disks were generally published only with annual volumes. Intermediate issues of data therefore – again – had to be manually input. Further, any revised data were not always published in an identifiable way.

At one stage in my career, I was responsible for the production of historical volumes of statistics for a particular period of area organisation of the country.[8] The idea for compiling the data series was simple enough: start with the latest volumes of subject-specific statistics; decide which series to include and start extraction. That produced the data for the last 5 years. Then take the penultimate volume of the published statistics for the one year at the start of the series; and then the antepenultimate volume for another year etc. back through history until the full run of 22 years had been completed. Before sending to the printers, each topic section was shown to the relevant statistician for confirmation of the extracted data. Several reported that some data were not the latest for all of the periods shown. On investigation, it became clear that some revisions to the data had not been published!

Then along came the Internet. The first data I put onto the Internet was in 1999. Looking back, it was progress in data dissemination but both the Internet and associated tools for the user had yet to be developed fully. Many organisations put data releases onto the Internet in very user-unfriendly formats, such as PDF or locked fixed tables. To extract data from a PDF output, the user had to either purchase a full version of the product to unpack the PDF file or manually input the data since copying a table and pasting into Excel resulted in each line of the table being put into the first cell of each row. One could parse the data splitting the data on a space – unless the table was from a country that used the space as a thousands separator!

Worse, when reviewing a website of a European national statistical institute to find the population figure, a hyperlink was followed to the yearbook of the country – and resulted in the downloading of a PDF file with a size over 18 Mb. The first time undertaking this task was in the early days of the Internet using a dial up line of 14.4k bits per second: needless to say, the task was abandoned! The exercise repeated around 2010 and the yearbook was still to be found in a single file in PDF format – but this time, using a 30 Mb fibre optic link connection, the download was completed. Still the information was not easy to extract from the file. Whilst it could be an advantage for a user to download the whole of a large publication, it is more usual that they require just one or two figures. It is not too great a challenge to split the book up into chapters and give the user the option of downloading the whole book or just an individual chapter.

The PDF format was the main one used by many national statistical institutes for press releases and other documents, including yearbooks, until around 2005. The first versions of these had all of the tables within the PDF and not accessible meaning a great deal of retyping of data. Later implementations of PDF files could include links to the tables in MS Excel format which then obviated the need for retyping.

Before this, in Wales, a different approach had been adopted which was much more user-friendly. Its basis was that the user generally did not want the detail given in press releases but, in the main, only the key points of interest in the newly released statistics. The principle was that the following information should appear on one screen at the same time:

- The title of the release;

- Date of release;

- Link to the full release in a MS Word document;

- Link to the tables of data;

- Up to five key points from the data – each of no more than one short sentence;

- The contact name, telephone number of the responsible statistician;

- The email address for enquiries on the release and

- The date of the next release of similar statistics.

Since the original design, some changes have been made and a page now looks as that given in Figure 5.2. The format of the press release has changed from an MS Word document to a PDF document. Where data are available separately, however, links are provided at the bottom of the release to the relevant database.

Figure 5.2   Extract of a web dissemination page.[9]

Another example of this principle is found on the Office for National Statistics website, as in Figure 5.3. This example does put the majority of the important information on one page but the link to the Statistical Bulletin takes the user to another intermediate page before linking to the actual bulletin. Here, there is a clear link though to reference tables of data that are available for download.

Figure 5.3   Example of a web dissemination page relating to a press release.[10]

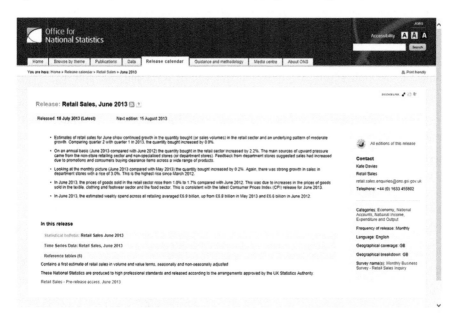

5.2   Statistics on CD-ROMs

In the late 1990s, many national statistical institutes sought to disseminate statistics on CD-ROMs. These varied from the practically unusable to some very well organised and usable products. The Czech Republic produced effective versions of their year book with simple access to the data: later versions provided different formats of output to meet user preferences. The Australian Bureau of Statistics also produced versions of their yearbook on CD-ROM but the early editions were difficult to extract the data from since most of the presentation was in PDF format.

Another issue with PDF files is navigation. Modern versions of the PDF creator tool do allow bookmarks to be saved with the files so that users of large PDF files can navigate through the file's structure to the part they want rather than having to search from beginning to end to see what is required. Sadly, many offices ignore this facility and produce large files without user helps.

Some offices decided to use CD-ROMs for the dissemination of other collections of data since the production and distribution of a CD-ROM was much cheaper than preparing, printing and distributing a paper version. One, issued in 1998, by the United Kingdom's Office for National Statistics gathered statistics for many health topics over 100 years and disseminated the data in Excel tables.[11] Notwithstanding that the office had a very comprehensive style guide for the dissemination of data, many of the tables did not follow the guide with some columns of data centred. One particular table that sticks in my memory presented the data on infectious diseases. Three of the columns presented data in thousands and one in units. Table 5.1 presents an extract of the table showing the first five and last 12 years of data.

Table 5.1   Mortality from smallpox, malaria, enteric fever and tuberculosis.

Year	Smallpox (thousands)	Malaria/Ague/ Remittent fever	Enteric fever (thousands)	Tuberculosis (all forms) (thousands)
1838	16.268	226		68.540
1839	9.131	231		69.165
1840	10.434	381		70.279
1841	6.386	284		69.824
1842	2.715	305		69.926
.. .. ..				
1979		6	0.002	0.613
1980	0.001	8	0.001	0.605
1981		2	0.002	0.557
1982		10	0.003	0.564
1983		7	0.001	0.467
1984		4	0.000	0.490
1985		5	0.001	0.523
1986		3	0.000	0.471
1987		7	0.002	0.430
1988		6	0.001	0.478
1989		4	0.000	0.390
1990		3	0.002	0.422

The largest number in the 100 years was only of the order of tens of thousands and the lowest in single units: so the data should have been given in actual numbers.

This table has:

- Numbers centred in columns;

- Missing numbers or symbols to represent nil or not available in the smallpox column for 1979 and from 1981 and the start of Enteric fever column;

- Differential number presentation (thousands and units) and

- Nonsensical portrayal of unit data (e.g. 0.002 thousands = 2!)

So a better presentation would be in Table 5.2 which follows the style guide of the originating organisation.

Table 5.2   Mortality from smallpox, malaria, enteric fever and tuberculosis.

Year	Smallpox	Malaria/Ague/ Remittent fever	Enteric fever	Tuberculosis (all forms)
1838	16,268	226	..	68,540
1839	9,131	231	..	69,165
1840	10,434	381	..	70,279
1841	6,386	284	..	69,824
1842	2,715	305	..	69,926
.. .. ..				
1979	–	6	2	613
1980	1	8	1	605
1981	–	2	2	557
1982	–	10	3	564
1983	–	7	1	467
1984	–	4	–	490
1985	–	5	1	523
1986	–	3	–	471
1987	–	7	2	430
1988	–	6	1	478
1989	–	4	–	390
1990	–	3	2	422

– = nil or half the final digit shown

.. = Not available

Microsoft themselves were early leaders in dissemination and brought out in 1993 a digital multimedia encyclopaedia on CD-ROM named Encarta. The principle employed was that of hyperlinking both to the different aspects of the information (what we now call navigation) but also to explanations of words in the text. This latter use is similar to the Wiki system of information dissemination.

These design principles are rarely used in statistics. However, in the last couple of years, such a system, Statistics Explained,[12] has been developed by the Eurostat, for the dissemination of meta-information about statistics: a sample page on consumer prices is given in Figure 5.4. It is essentially a closed Wiki system with only members of staff allowed to update information – and, at the time of writing, is the best such system of disseminating meta-information on statistics.

Figure 5.4   A page from Eurostat's Statistics Explained.

The navigation is clear, in a single colour and offers users explanations of terms, methods and context. Also, links are provided to the databases and data visualisations. This is similar to the Encarta model in that all of the meta-information is gathered in one place. Often the user would prefer, when looking at data tables, the possibility of identifying a word or phrase in the surrounding row or column text that they do not understand and being able to see that it is hyperlinked. This hyperlink can then lead to a pop-up of definition and/or explanation. Some statistical dissemination products have the facility to put information symbols in cells, against text or numbers which are links to glossary items or footnotes. This should be standard, assisting all users.

5.3   Data on the Internet

The basic principle of data on the Internet is accessibility. If data are difficult to access through the now simple and universally available Internet, it is the dissemination that is wrong. This decade is seeing an explosion of different types of devices that can access the Internet, from mobile phones through iPads to netbooks, portable computers and desktop computers. Some of these bring consequential challenges to disseminators because of the small screens.

The first significant trial use of statistics dissemination with wireless access protocol for mobile phones was by Statistics Finland. This was not successful as, at the time, mobile phones had very limited, non-graphic screens to display data and few had significant Internet access. What a change has been effected in the last 10 years! The majority of mobile phones now are graphics-based and come with reasonable Internet access.

Over this same period, the whole philosophy of statistics dissemination has also changed. Now, instead of the producer of statistics designing large and complicated tables and insisting that the user has to download masses of data, the user has the power to extract (dice and slice) data from large data sets and prepare just the amount of information they require. Some dissemination programs also provide the user with the ability to add derived rows or columns to the data table being extracted. So, instead of having to extract a table to Excel and do the derivation of sub-totals, totals, ratios or percentages, it can all be done within the one extraction process.

PC-Axis,[13] developed by the Statistics Sweden, is one good example of this. This program has now been implemented in an Internet version, PX-Web, by many national statistical institutes for dissemination of their statistics. Data are stored in multidimensional databases of up to six dimensions. The user can select a dataset and is then shown the various dimensions (see Figure 5.5). The next stage is for the user

Figure 5.5   Sample selection screen from PX-Web.[14]

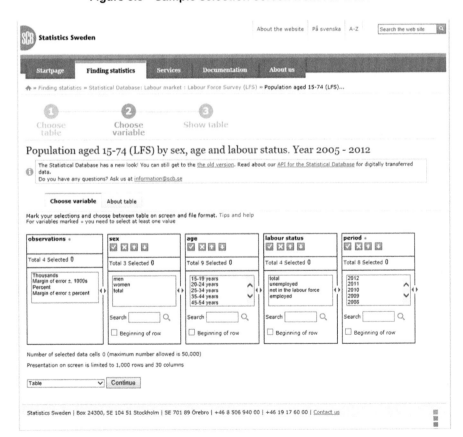

to select which categories of each variable are needed: the program allows for all of the categories to be selected or the user can choose from the list of categories within each variable list. Once all dimensions have been defined, the user requests the table and it appears on screen with another set of options – how to harvest the table for onward use or with an option to display the data in a chart.

Some data do not lend themselves to display in a chart but the program will allow production: it is up to the user to decide if the chart is appropriate. In some implementations of the program, mapping is also possible. A similar extraction program was developed as a front end to Oracle databases by the Food and Agriculture Organisation of the UN[15] – which makes extraction a simple task for all without the requirement to be an excellent user of Oracle databases.

Another example of a good dissemination system is StatLine,[16] developed by the Statistics Netherlands,[17] which handles all statistical outputs and allows manipulation of data as well as mapping. This latter complete dissemination system also includes a very good search tool. If one searches for a bicycle, the program returns a list of outputs that has information on bicycles including reports, press releases, tables and maps.

Other systems clearly have not had the strong hand of statisticians assisting with design. Some of these place row totals to the left of tables and column totals at the top of columns – as it is easier to program that outcome.

But, back to basics, many still disseminate tables in Excel workbooks. This is still a good method and the extracted table can be easily manipulated to produce a table required by the user. For those looking to use this means – or those who already do – it may be worth mentioning a couple of principles that should be observed when disseminating information in this way. First, any spreadsheet should always be saved with the cursor in cell A1: this ensures that the user, on opening the spreadsheet, can see the table title. One of the better presentations of spreadsheets on the Internet is given by the Statistical Office of Luxembourg: they display an office logo at the top of the sheets[18] as in Figure 5.6 so that, in printing, one knows the source. Second, when many sheets are organised in the workbook, some help needs to be given to the user so that they can find the relevant sheet easily. Some name the sheet tabs in the workbook clearly and effectively but naming the tabs T1, T2, T3a, T3b etc. is not that helpful! If the titles are long or the workbook contains many sheets, it is probably simpler if the first sheet in the workbook has a list of the tables in, essentially, a contents page where each table title is hyperlinked to the relevant table. Again for both the index sheet and all sheets within the workbook, the user should be presented with the cell A1 visible in the top left of their screen.

Figure 5.6 Example of a headed spreadsheet.

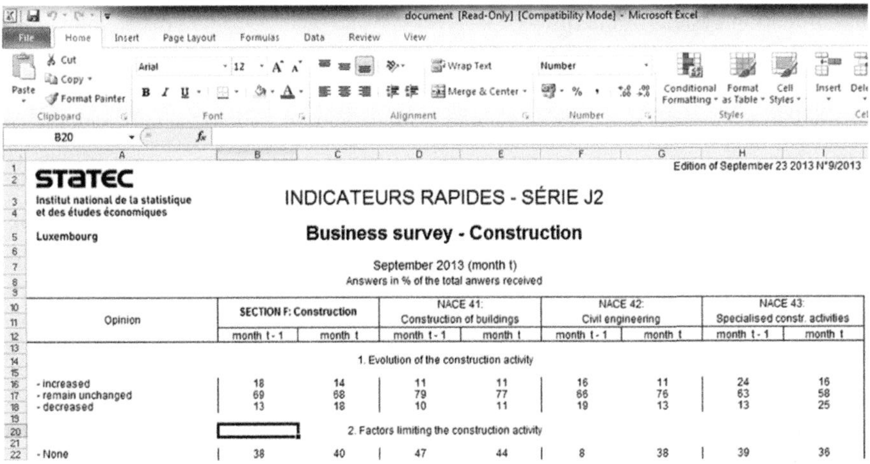

STATEC
Institut national de la statistique et des études économiques
Luxembourg

Edition of September 23 2013 N°9/2013

INDICATEURS RAPIDES - SÉRIE J2

Business survey - Construction

September 2013 (month t)
Answers in % of the total anwers received

Opinion	SECTION F: Construction		NACE 41: Construction of buildings		NACE 42: Civil engineering		NACE 43: Specialised constr. activities	
	month t - 1	month t	month t - 1	month t	month t - 1	month t	month t - 1	month t
1. Evolution of the construction activity								
- increased	18	14	11	11	16	11	24	16
- remain unchanged	69	68	79	77	66	76	63	58
- decreased	13	18	10	11	19	13	13	25
2. Factors limiting the construction activity								
- None	38	40	47	44	8	38	39	36

Principle 5.1: Ensure the cursor is in cell A1 before saving Excel tables to be accessed through the Internet.

Principle 5.2: If saving more than one Excel table to be accessed through the Internet, provide an index sheet and save the file with the cursor in cell A1 of the index sheet. Each element of the index can then be hyperlinked to the relevant sheet.

Many national statistical institutes produce a significant number of volumes, each containing many tables. Thinking of how a user would access a yearbook presented in paper form led to an experimental presentation on the Internet. The Digest of Welsh Statistics moved from hot metal typesetting in 1980 to photolithographic production in 1981 and onto the Internet, with continuing paper publishing for 20 years. The experimental presentation was based on the way a user would try to find a set of statistics – via the content pages at the front of the book or through the index at the back. For the Internet version, it was not uploaded as a single file but had one introductory page with a brief chapter summary page. The contents page linked to a page designed as in the paper edition but each table title was itself linked directly to an Excel version of the table. If the user chose to open a chapter from the chapter summary page, one could choose from downloading the whole in MS Word or Excel format or selecting one of the hyperlinked individual tables. If using the index, where one would just find the table numbers for particular subjects, in these editions on the Internet, each of the table numbers shown against the keywords in the index was hyperlinked to the Excel version of the table. This presentation only lasted two editions as it was very time consuming to effect. Since 2003, it has not been produced either on paper or on the Internet but most of the data are available through the dissemination system of the office.

> Principle 5.3: Organise collections of spreadsheets effectively to allow easy access.

In addition to the organisation of workbooks, it is essential to pay attention to the presentation of the data themselves especially in their formatting. Since formatting of data within Excel or similar programs does not destroy the basic data, it may be appropriate to round some of the data in the tables. Users can always then change the formatting to suit their own purposes, if necessary.

> Principle 5.4: Format data in Internet-published spreadsheets to appropriate standards.

Some Internet versions of press releases have linked data tables. It should not be necessary to note that these linked tables should be appropriately structured with all of the necessary elements of a table present. Issues identified in such tables in the past year (when doing research for the Surfing with Ed[19] articles) included

- tables with data in millions but no table descriptor;
- data columns centred, not right-justified;
- inconsistent symbols for decimal and thousands separators and
- absence of sources.

These outputs do not conform to their organisation's standards for dissemination – but should do.

> Principle 5.5: Maintain dissemination standards in all Internet data outputs.

5.4 Charts on the Internet

Obviously any chart can be put onto the internet as a picture. In some implementations, users are also given links to the underlying data. During some reviews of National Statistical Institute websites, I have come across such links, followed them and found that the data are not in a properly formatted table: some data did not have descriptors (such as thousands or millions)! But this is not really using the power of the Internet to its fullest extent. Indeed, statisticians have been relatively slow to adopt the possibilities offered by today's Internet.

Gone are the days when what is shown on a piece of paper is replicated on the screen. Just as tables are now able to be diced and sliced dynamically to give the user just what they want, so the power of the Internet is being harnessed to give much more than a simple picture. One of the earliest implementations of a dynamic chart in the United Kingdom was a population pyramid using Scalable Vector Graphics (SVG) in around 2003. Such charts were common in the paper publications on population structure and change. The difficulty for users was that the SVG viewer software was a requirement for viewing and using the chart and had to be installed on the user's computer: some organisations did not allow staff to download software – including the SVG viewer. Later implementations are based on Flash which is available on most computers at initial set up and updated centrally for office-wide systems as accepted software.

So what is different in the Internet presentation? First, the chart is not a static chart like a picture but can be made to change across another variable, often time; secondly, the data behind the chart are used dynamically to show changes – in this case, over time – and can be displayed. Other presentations also allow investigation of the data in the chart and accessing of the individual data values. Figures 5.7–5.9 show screens from a demonstration population pyramid.[20]

Figure 5.7   Opening view of Population Pyramid.

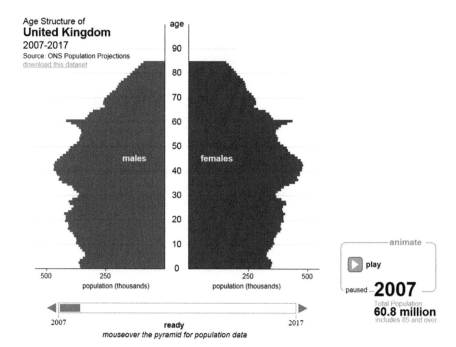

Now, apart from the basic chart, we have a slide bar beneath the chart, an instruction below that to show data for a particular age and a 'play' button to the bottom right. Clicking on the 'play' button, displays the 2008 version of the pyramid, then 2009, etc. Note that the total population figure also changes as the different years are shown.

The instruction below the slide bar tells the user to move the mouse over the pyramid to get the data for each age group as in Figure 5.8.

Figure 5.8   Pyramid with data shown for those aged 60.

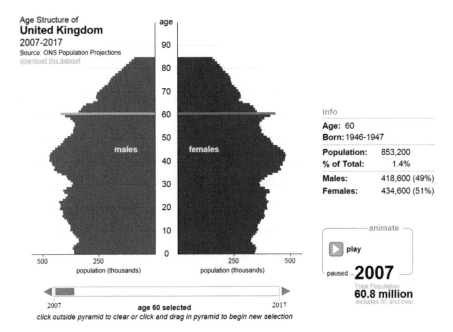

Clicking on the play button now changes the pyramid over time as before but also changes the displayed data for those aged 60 through the years. Note also that the instruction below the slide bar has also changed. The user does not just have to select one age but can select multiple ages. Once these have been selected, as shown in Figure 5.9, clicking on the play button shows the data for the combined ages. The version illustrated here is a demonstration sample: the full version of this pyramid showing many more years of data is available on the Office for National Statistics website.[21] In the full version, try selecting those aged 80–84 and see how the total population of these ages is expected to change over the period – very worrying for the providers of care for this age group and those who have to provide pensions.

Figure 5.9   Pyramid with data shown for those aged 80–84.

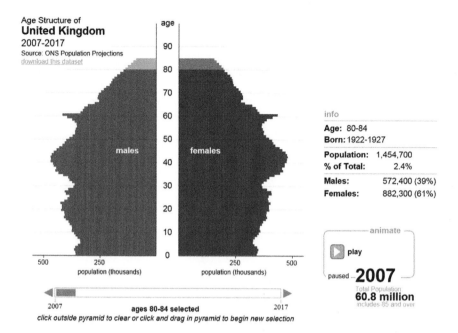

Even in the simplified version used to illustrate the population pyramid here, the user has to assess the changes across 11 charts. This is not a simple task and those developing data visualisation presentations are beginning to find out that the user requires some help to see what is happening. For the pyramid example, it is the provision of several features:

- An outline of the starting position so that the user can see the changes to whatever time they want to examine;

- A speed selector button – from slow to fast and

- The play/pause button to get an idea of changes to certain dates within the overall range.

Hans Rosling developed the Trendalyzer software which again allows the user to see changes over variables over time. The first public version of this software I saw was based on the income per person (inflation adjusted) against life expectancy for a vast number of countries of the world: data from 1975 to 2004 were available. The web page opened, Figure 5.10, with the first year shown and each country had a coloured bubble on the chart. The size of the country's bubble was proportional to its population and the colour of the bubble designated which area of the world the country was in.

Figure 5.10   Opening screen of old version of Gapminder World.[22]

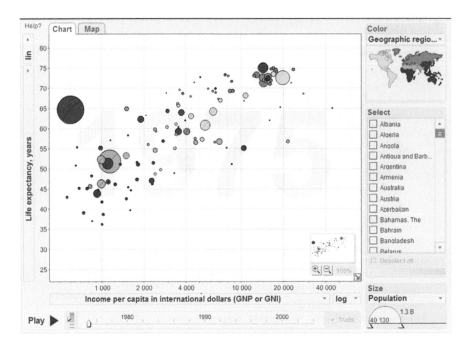

The user could select specific countries to examine from a list to the right of the chart with a slide bar to the right of the chart. For this example, China, South Africa and Rwanda were chosen. The display then highlights the chosen countries' bubbles and reduces the intensity of the others, as in Figure 5.11.

Figure 5.11   Chart with three countries selected.

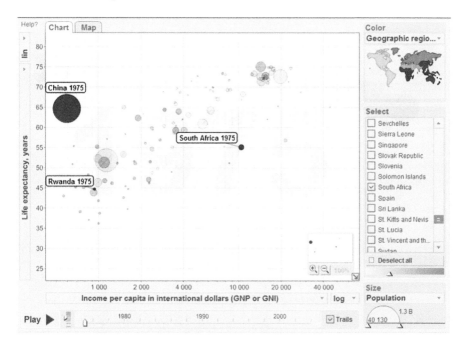

The user then clicks on the Play button and the story or the data is displayed, as in Figure 5.12. It is clear from this figure that some change in direction occurs in both the series for Rwanda and South Africa: when the mouse hovers over these turning points, the year is shown together with the data values. Explanations can then be sought for the reasons for such changes in the data. The series over this period for China does not have any turning points but shows a relatively small rise in life expectancy but a very large increase in income per capita. Such a tool allows extensive and easy investigation into data in a way that has not been as easily possible previously. However, with the vast amounts of data available, it is necessary to apply a critical understanding to the selection of data in order to gain the appropriate message.

In Figure 5.12 the data for each of the years is displayed in trails of the selected series over time (allowing the user to compare trends over time for different countries). This facility can be turned off when too many countries have been selected – otherwise the whole chart is filled with trails that cannot be easily observed.

Figure 5.12   Chart with visual of three countries over time.

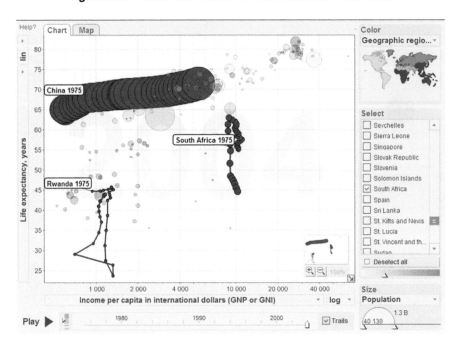

The latest version[23] includes

- a link to user instructions;
- more and longer series of data;
- a map key and
- dynamic data provision (if the user hovers over a specific bubble on the chart, the country name and data values are displayed and, in addition, the map to the top right of the display indicates the region of the world that the country is classified to).

These pages from Gapminder only look at the two base series of data. Many different series can be selected from drop-down menus accessed by clicking on the arrows to the right of the variable names shown – that is, on either axis. Some of the variables are only available through a 'For advanced user' link to aid the user. It is up to the user to determine whether any real relationship exists between the variables or not. Further, to the right of the axis labels, another option drop-down box allows a change of scale type, from linear to logarithmic.

The Trendanalyzer software can be utilised to create dynamic charts with the user's own data. Some are published on the Internet, for example, Statistics South Africa[24] (select Interactive Graphs from the navigation links at the bottom of the page) and the

Greater London Authority.[25] Other types of data visualisation are being introduced by many national statistical institutes, for example, the US Census Bureau's gallery of interactive data visualisations,[26] the CBS Netherlands' interactive infographics[27] and the Office for National Statistics (UK) Interactive Content.[28]

Data visualisation applications are not limited to charts. The American Community Survey[29] and the American Census[30] have used data maps extensively in summarising results. The Data Unit in Wales[31] uses maps as part of its dashboard for local authorities named InfoBase Cymru:[32] an example of this is given in Figure 5.13. This application links data, chart, map and historical information and presents all on one page. The user can choose different data sets, some with more area detail: in those cases, the user can zoom in on the map portion and move it within that part of the screen to investigate the data for any characteristic or outliers.

Figure 5.13   Example page from InfoBase Cymru.

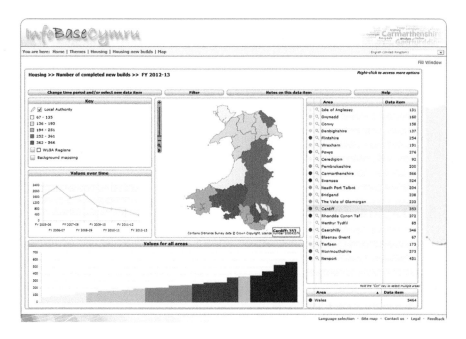

Just before taking the screenshot above, the mouse was hovering over Cardiff – and the appropriate data value is displayed – on the map. Consequently, the area on the map, the data for Cardiff on the right and Cardiff's bar in the bar chart were all highlighted in green. In addition, the time series of data was plotted on the graph on the left. It is possible to highlight more than one area at once to be able to compare data items: hold CTRL whilst clicking on the relevant areas of the map or the data lines in the table.

Much work is going on as this book is written to develop data visualisation techniques. Issues are being identified and addressed. For example, should one wish to look at the United Kingdom's population pyramid example from the live edition,[33] the user has to visually manage 115 different pictures and seek to identify differences over time. The pause button can allow appreciation of a point in time but it is still difficult to assess the change from the start. Adding an outline to the chart at the first year (or any year would be better) is in development: this would enable the differencing to be seen clearly. It has also been discovered that a slower change (altering the time change slide bar on this graphic) is more helpful to the visual perception.

Data-driven documents – known as D^3 or D3 – are the latest development of data visualisation. These visualisations allow the user direct power to change the presentations online and without requiring additional software. D^3 visualisations also load onto the user's screen faster and can change quicker as well.

Effective data visualisation allows the user to explore the data live on their computer screen. Such exploration then facilitates the development of messages from the data and allows the user to understand better the relationships within the datasets. Ease of use is paramount and the majority of applications available are easily understood and used. Early application of one product showed the data in a table left-justified and to different numbers of decimal places – thus obscuring the visual scan of the data to identify the largest or smallest value. Throughout visualisations, it is essential that the basic principles of data presentation are followed.

> Principle 5.6: Ensure basic chart principles (see summary at the end of Chapter 3) are followed for dynamic charts and data visualisations.

5.5   Text on the Internet

The way users want to access the written information on the Internet is very different from what used to happen with, say, reports of surveys in book form. It is very much a point-and-click access with very little reading of dense text. So the producers have to consider their output and identify whether changes need to be made to the standard styles of output before transferring the output from paper base to Internet base.

The examples of Internet pages about the release of new statistics in Section 5.1 go a little way to the change in concept which is basically an inverted style of writing:

- Headline
- Summary in bullet points
- Detail including summary tables
- Links to complete reference data

But most of the output on the Internet seen today does not fully utilise the power of the web to meet the expectations of the Internet generation. The use of links within statistical output documents is poor both for skipping around the document and for explanations of technical terms. Further, the amount of text that some try to squeeze onto a page does not follow the user's demand for quick access. For announcements of press releases, it is not good enough simply to say:

Transport, September 2013 – final data

The key message from the release should be given: if the advice to identify key messages and sort them into significance order has been followed, the most important message can be displayed along with the above release title as:

Transport, September 2013 – final data

In September 2013, 18.4% more passengers were carried in urban scheduled transport, compared to September 2012.

Not putting these key messages on the page of press releases forces the user to open the release itself to try to find the key message. If the key message is displayed in the first instance, the user has the option to dig deeper or to move in to something else. This does mean that the message displayed should be the key one, brief, relevant and appropriate.

The Internet user scans a page to find the keywords that interest them and then they follow the appropriate links. Clear navigation with simple headings and headlines is fundamental to facilitate easy use of websites. Embellishments of sites with flashing graphics, large pictures and non-translatable text in graphics do not enhance usability. Some guidance on features of websites – both statistical and others – that are good and bad is available as a checklist on my website.[34]

Principle 5.7: Ensure material published to the Internet is appropriately designed for Internet use.

Summary of principles: Data presentation on the Internet

Principle 5.1: Ensure the cursor is in cell A1 before saving Excel tables to be accessed through the Internet.

Principle 5.2: If saving more than one Excel table to be accessed through the Internet, provide an index sheet and save the file with the cursor in cell A1 of the index sheet. Each element of the index can then be hyperlinked to the relevant sheet.

Principle 5.3: Organise collections of spreadsheets effectively to allow easy access.

Principle 5.4: Format data in Internet-published spreadsheets to appropriate standards.

Principle 5.5: Maintain dissemination standards in all Internet data outputs.

Principle 5.6: Ensure basic chart principles (see summary at the end of Chapter 3) are followed for dynamic charts and data visualisations.

Principle 5.7: Ensure material published to the Internet is appropriately designed for Internet use.

Notes

1. http://archive.org/web/

2. New Zealand: http://natlib.govt.nz/collections/a-z/new-zealand-web-archive?search%5Bpath%5D=items&search%5Btext%5D=web+archive
UK: http://www.bl.uk/aboutus/stratpolprog/digi/webarch/

3. http://www.nationalarchives.gov.uk/webarchive/

4. US Census: https://archive.org/details/us_census
UK Data Archive: http://www.data-archive.ac.uk/

5. British Labour Statistics: Historical Abstract 1886–1968, HMSO, London, 1971.
John Williams (1985). *Digest of Welsh Historical Statistics Volume 1*, Cardiff: Welsh Office. ISBN 0 86348 248 1.
John Williams (1985). *Digest of Welsh Historical Statistics Volume 2*, Cardiff: Welsh Office. ISBN 0 86348 249 X.
Historical Statistics of the United States: Colonial Times to 1970, US Census Bureau,
Part 1: http://www2.census.gov/prod2/statcomp/documents/CT1970p1-01.pdf
Part 2: http://www2.census.gov/prod2/statcomp/documents/CT1970p2-01.pdf

6. http://wales.gov.uk/topics/statistics/publications/dwhs1700-1974/?lang=en

7. http://discover.ukdataservice.ac.uk/catalogue?sn=4095

8. Williams, L.J. (1998). *Digest of Welsh Historical Statistics, 1974 to 1996*, Cardiff : Welsh Office. ISBN 0 7504 2299 8.

9. http://wales.gov.uk/topics/statistics/headlines/environment2013/energy-generation-consumption-2011/?lang=en
Reproduced by permission of the Welsh Government.

10. http://www.ons.gov.uk/ons/rel/rsi/retail-sales/june-2013/index.html
    Reproduced under the Open Government Licence.

11. 1998. *The Health of Adult Britain 1841 to 1994*, London: Office for National Statistics.
    ISBN 1 85774 294 X

12. Statistics Explained published on the Internet by Eurostat at
    http://epp.eurostat.ec.europa.eu/statistics_explained/index.php/Main_Page
    Used with permission of Eurostat.

13. http://www.scb.se/Pages/List____313989.aspx

14. Reproduced by permission of Statistics Sweden.

15. http://faostat3.fao.org/faostat-gateway/go/to/home/E

16. http://statline.cbs.nl/statweb/?LA=en

17. http://www.cbs.nl/en-GB/

18. http://www.statistiques.public.lu/stat/TableViewer/document.aspx?ReportId=1345&IF_
    Language=eng&MainTheme=4&FldrName=4
    Reproduced with permission of Statistics Luxembourg.

19. https://surfingwithed.wordpress.com

20. Reproduced from Office of National Statistics under the Open Government Licence, UK

21. Population pyramid for UK population projections is at
    http://www.ons.gov.uk/ons/interactive/uk-population-pyramid—dvc1/index.html

22. The three charts are free material from www.gapminder.org

23. http://www.gapminder.org/world/

24. http://beta2.statssa.gov.za/

25. An example at
    http://data.london.gov.uk/visualisations/charts/fol11-poverty-charts.html

26. http://www.census.gov/dataviz/

27. http://www.cbs.nl/en-GB/menu/publicaties/webpublicaties/interactief/default.htm

28. http://www.ons.gov.uk/ons/interactive/index.html

29. https://www.census.gov/acs/www/

30. http://flowsmapper.geo.census.gov/flowsmapper/flowsmapper.html

31. http://www.dataunitwales.gov.uk/

32. http://www.infobasecymru.net/IAS/eng

33. http://www.ons.gov.uk/ons/interactive/uk-population-pyramid—dvc1/index.html

34. https://surfingwithed.wordpress.com/web-design-guidance/

Printed and bound by CPI Group (UK) Ltd, Croydon, CR0 4YY

27/10/2024

14580209-0001